ENERGY SCIENCE, ENGINEERING AND TECHNOLOGY

PHOTOVOLTAICS

SYNTHESIS, APPLICATIONS AND EMERGING TECHNOLOGIES

ENERGY SCIENCE, ENGINEERING AND TECHNOLOGY

Additional books in this series can be found on Nova's website under the Series tab.

Additional e-books in this series can be found on Nova's website under the e-book tab.

ENERGY SCIENCE, ENGINEERING AND TECHNOLOGY

PHOTOVOLTAICS

SYNTHESIS, APPLICATIONS AND EMERGING TECHNOLOGIES

MONIQUE A. GILL
EDITOR

nova publishers
New York

For permission to use material from this book please contact us:
Telephone 631-231-7269; Fax 631-231-8175
Web Site: http://www.novapublishers.com

Additional color graphics may be available in the e-book version of this book.

LIBRARY OF CONGRESS CATALOGING-IN-PUBLICATION DATA

Photovoltaics : synthesis, applications and emerging technologies / editor, Monique A. Gill.
pages cm. -- (Energy science, engineering and technology)
Includes index.
ISBN 978-1-63117-843-6 (hardcover)
1. Photovoltaic power generation. 2. Photovoltaic cells. I. Gill, Monique A., editor of compilation.
TK1087.P469 2014
621.3815'42--dc23

2014013788

Published by Nova Science Publishers, Inc. † New York

CONTENTS

PREFACE

Increasing demands on energy has necessitated the development of large-scale, low-cost, renewable energy. Solar energy is highly attractive and extremely promising as the energy reaching the earth from the sun is orders of magnitude larger than today's global energy consumption and is quiet, requires low maintenance, has no restrictions on size, can be located close to the point on consumption, and is a clean, direct and efficient mode of energy conversion. The challenge lies in making solar power economically competitive while working with the materials readily available in abundance along with all their inherent constraints, also using those materials as economically as possible. Novel materials and structures are constantly being investigated for possible application in the area of photovoltaics (PVs). This book begins by discussing nanostructure-based PVs. It then continues to provide information on several topics that include the environmental impacts from end-of-life disposal of PVs; photocured polymer electrolyte membranes for dye-sensitized solar cells; economic analysis of support policies in PV systems; indoor PVs; effect of urban density on PV performance; and PV effects in nanoscale ferroelectrics.

Increasing demands on energy has necessitated the development of large-scale, low-cost, renewable energy. Solar energy is highly attractive and extremely promising as the energy reaching the earth from the sun is orders of magnitude larger than today's global energy consumption and is quiet, requires low maintenance, has no restrictions on size, can be located close to the point on consumption, and is a clean, direct and efficient mode of energy conversion. The challenge lies in making solar power economically competitive while working with the materials readily available in abundance along with all their inherent constraints, also using those materials as economically as possible. Novel materials and structures are constantly being investigated for possible application in the area of photovoltaics (PVs). Lately, various nanoparticle-polymer-based solar cells have demonstrated reasonable power conversion efficiencies (PCE), about 9%. Theoretical predictions suggest that higher efficiencies are achievable through nanocrystal-based approaches due to the outstanding optoelectronic properties and multi-exciton generation capabilities of semiconductor nanocrystals (NCs). Hence, these hybrid solar cells have the potential to provide cost-effective, high-efficiency solutions for PV, although significant challenges need to be overcome.

Chapter 1 discusses several approaches in pursuit of efficient light absorption in materials. Group-IV materials such as silicon (Si) and germanium (Ge) are desirable for optoelectronic applications due to their high abundance, cost-effectiveness, and high compatibility with the existing technology but as they are indirect band gap materials, they inherently have poor optoelectronic properties at room temperature. Approaches such as point defects, optically excited Raman lasers, quantum confinement in Si NCs, utilizing rare earth dopants, bonding of III-V semiconductor structures onto Si chips, integration of active conjugated-polymers with Si, synthesis of tensile-strained semiconductor heterostructures and NCs to modify their band structure and obtain favorable optical properties which result in improved carrier mobilities have been tried to address the challenge.

Band engineered nanoparticles (NPs) and NCs provide improvements in the primary photoconversion process by increasing the energy absorbed from the electromagnetic spectrum that may otherwise be lost as heat in bulk semiconductors and thin films. Such NS may be embedded in a suitable matrix for application in PV and could be a potential solution to surpass the ~33% Shockley-Queisser (SQ) limit for PCE, leading us towards viable "third generation" PV devices. This approach also offers great flexibility, allowing us to combine various materials to maximize the range of wavelengths absorbed by the materials which could lead to higher efficiencies for PVs. Also, various nanostructured materials are being experimented for the contact layers. The following sections discuss the synthesis and characterization of the structural and optoelectronic properties of such nanostructures (NSs) and their advantages for high efficiency PV.

In Chapter 2 is presented the global solar photovoltaic (PV) deployment that enjoyed exponential growth in the past decade. As the industry grows, large future end-of-life PV waste volumes will need to be managed. It is likely that recycling will be a dominant strategy as in the case of the European Union, where end-of-life collection and recycling will be mandatory beginning 2014 under the recast Waste from Electrical and Electronic Equipment Directive. However, PV deployment is globalizing beyond the European Union into emerging markets. Given the presence of various metals in PV modules, some stakeholders have raised concerns regarding potential environmental impacts if PV modules are disposed of in unlined landfills instead of sanitary landfills or instead of being recycled. Potential adverse health or environmental impacts associated with disposal of utility-scale (25 MWac) volumes of PV modules in an unlined landfill were evaluated by fate and transport modeling using U.S. EPA Delisting Risk Assessment Software (DRAS). Estimated surface and groundwater media concentrations of Cd and Pb, and associated cancer risks and non-cancer hazards were below human health and ecological screening limits. The hazard index under basic landfill conditions was an order of magnitude lower than under acidic conditions, which is notable as landfills have predominantly neutral to slightly basic conditions over their lifetime. Estimated environmental concentrations of Pb from crystalline silicon PV were comparable to concentrations of Cd from cadmium telluride PV, indicating that responsible disposal is important for all PV technologies. Potential impacts may be further limited by use of sanitary landfills or with high value recycling, with the latter providing the important benefit of resource recovery. The DRAS default assumption is Toxicity Characteristic Leachate Procedure (TCLP) data for waste with a fragment size of 1 cm. Experimental crushing of cadmium telluride PV modules in landfill conditions was conducted to evaluate the representativeness of the leachate data used in the DRAS modeling. On average approximately three-quarters of the crushed module fragments were larger than 1 cm, and 99% of module fragments were larger than 0.1 mm. The glass-laminate-glass bond of individual broken module fragments was maintained, and when analyzed without further size reduction, the landfill fragments are non-hazardous based on TCLP and Soluble Threshold Limit Concentration (STLC) testing.

The photovoltaic market is currently dominated by silicon technology, but emerging devices such as Dye-Sensitized Solar Cells (DSSCs) are drawing increasing attention, due to their easy fabrication, low cost and high conversion efficiency. However, one of the major problems limiting the long-term stability of these devices and their subsequent large-scale industrial production is the volatilization of the liquid electrolytes traditionally employed. Furthermore, a liquid also presents technological drawbacks with respect to its effective encapsulation in an electronic device.

In order to solve this problem and improve the technological perspectives of DSSCs, many studies have been focused on the preparation of quasi-solid electrolytes, in which a polymer network is exploited to effectively retain the redox mediator. In Chapter 3, the photoinduced polymerization has attracted large attention, demonstrating to be the most effective preparation method for these polymer electrolytes. In fact, among other advantages, it is a rapid, economic, functional and environmentally friendly process, besides being easily transferable to an industrial scale.

In this chapter, the fabrication and characterization of free-standing UV-cured polymer electrolyte membranes and their integration in quasi-solid DSSCs is reported. In particular, the first part of the chapter describes the dye-sensitized solar cell working principles and the photopolymerization process. In the second part, devoted to the new electrolytes, the discussion concerns the selection of peculiar monomers useful to obtain a specific 3D network, the development of smart fillers used to selectively improve the photovoltaic parameters, and the physico-chemical correlation between charge transport properties and polymer backbone architectures.

The development of the photovoltaic (PV) sector in the last decade was spurred by the implementation of various support strategies aimed at reducing the gap between PV energy cost and the energy price for conventional generation. The deployment of support policies has pushed the reduction of PV energy costs, but despite this PV is still not very competitive and its development still requires adequate support mechanisms, simple grid connection procedures, and so on.

Chapter 4 provides a technical-economic analysis of investments in PV systems, evaluating the positioning of the main European markets (Germany and Italy), with the purpose of highlighting the main differences in the implementation of support policies in the two top countries. Several case studies are examined, differentiated in relation to the rated power (from 3 kW to 1 MW) and the type of plant (ground-mounted and rooftop PV systems). The profitability indexes in terms of the payback period (PBP), net present value (NPV) and the internal rate of return (IRR) are evaluated throughout the reference period. The chapter is expected to be very useful for policy makers, energy producing industries and national Governments to estimate the profitability of PV investments in the two selected countries, predicting how the PV sector will evolve in each of the two EU states.

The uncertainty introduced by non-dispatchable generation, e.g. photovoltaic, affects the daily operation and planning of electric networks. To study this uncertainty, it is common to model the performance of photovoltaic units from a prediction of solar radiation. In Chapter 5 a method to predict this photovoltaic generation is proposed.

The purpose of this prediction is to obtain the output electrical power from the photovoltaic generator as a random variable. This random variable can be used in studies where it is considered appropriate to take into account the uncertainties, e.g. probabilistic load flow.

The clearness index and the fraction of diffuse radiation, which are considered random variables, are the basis of the model proposed. An expression that gives the global radiation on a sloped surface is obtained, depending on the input variables that are the clearness index and the fraction of diffuse radiation. After that, another expression that gives the output electrical power from the photovoltaic generator, also as a function of the input random variables, is reached.

Cumulants method is used to solve equation for output electrical power from a photovoltaic generator. To characterize the output random variable obtained, Gram-Charlier expansion is used.

The model proposed in this work takes into account the random nature of solar radiation, dependent on the geographical location where the solar panels are installed, and the hour and day of the year in which the simulation is performing. This approach allows determining the output electrical power from a photovoltaic generator at any place and hour as random variable.

Support schemes for solar PV have widely been adopted in the world. There is a voluminous literature on the theoretical and empirical analysis of different instruments to support renewable energy sources in genera and solar PV in particular. Chapter 6 has mostly focused on the effectiveness and efficiency criteria, although other criteria have also been included in some analyses. However, what is missing is a comprehensive framework which includes all relevant assessment criteria.

In addition, the assessment criteria have been treated in an isolated manner, although these criteria interact between each other.

Therefore, the aim of this paper is to close those gaps in the literature by providing a comprehensive framework which integrates all the relevant assessment criteria, explicitly considering their interactions. The results are deemed useful in order to undertake future multicriteria analyses of support schemes for solar PV.

The use of photovoltaic devices within buildings is also known as Indoor Photovoltaics (IPV). IPV has gained rising interest in the last years, when both the development of energy efficient electronics with an average power demand of a few micro watt and the introduction of narrow banded artificial light sources took place. This has opened new application fields for IPV, such as to power wireless sensor systems in intelligent buildings, with limiting efficiencies close to 50% for fluorescent lighting and even higher for LED spectra.

Chapter 7 provides the basic theory, recent progress, and future aspects of IPV devices. Aspects of IPV characterization are explained and referenced to methods from Standard Test Conditions following IEC 60904-3 Ed. 2 (2008). Typical indoor irradiance conditions and the resulting theoretical photovoltaic efficiency limits for ideal and real cells under these conditions are discussed. The calculated limits are compared to recent progress in research and industrial devices. The chapter closes with a discussion of current and future optimization strategies.

The development of sustainable materials in thin film form is a prerequisite for their potential deployment in future photovoltaic technologies. While cost, performance, and recently carbon footprint have been essential factors in materials selection, other equally important factors such as materials abundance (supply chain), chemical risk (toxicity), and recycling/reuse at end of life (cradle-to-cradle approach) have emerged as the new paradigm shift in the deployment of sustainable alternative energy technologies. Many concerted research efforts have been directed at the development of thin films that are able to guarantee optimal characteristics in terms of abundance, environmental compatibility, and photoactivity. Iron disulfide (pyrite) stands out as an excellent candidate for competitive photovoltaic technologies, as it meets all sustainability criteria in addition to its attractive technical capabilities. For over three decades, several investigations have been carried out to develop pyrite in thin film form and highlighted the potential of pyrite in future large-scale production of solar cells. Chapter 8 focuses on the review and discussion of different fabrication methods of pyrite thin films.

Buildings consume more than 40% of global energy use and are responsible for 33% of global greenhouse gas emissions. Hong Kong, being one of the most densely populated cities in the world, has buildings consuming over 90% of total electricity annually. Out of the various renewable energy sources, solar power using photovoltaic cells is considered the most suitable for wide-scale implementation in a densely populated city with a large number of high rises. To quantify the potential, a methodology was devised to determine the amount of Building Integrated Photovoltaic (BIPV) and Non-BIPV panels that may be installed on the rooftop and façades of selected existing buildings as well as in open space. With a database of building stock being established, a detailed site survey can then be conducted for the sampled buildings in order to determine and verify the feasible areas for PV installation. This methodology can be applied to other similar cities with a large number of high-rise buildings. In addition, the challenges and details of actual projects with different types of BIPV are described in Chapter 9. Operating measurements from the projects can then be used for design verification and further practical design considerations. Coupled with practical operating experience, this methodology can be used to evaluate the effect of urban density on photovoltaic (PV) performance.

Light matter interactions in nanoscale ferroelectric materials have received growing interest over the last years due to new developments in scientific instrumentation and novel materials that allow for the study of so far scarcely investigated and /or hidden nanoscale phenomena. Domain walls in complex oxides have especially been in focus opening new pathways for a number of possible applications. Recently an anomalous photovoltaic effect in ferroelectric $BiFeO_3$ thin films was reported that arises from a unique, new mechanism - namely, structurally driven steps of the electrostatic potential at nanometer-scale domain walls. In conventional solid-state photovoltaics, electron–hole pairs are created by light absorption in a semiconductor and separated by the electric field spanning a micrometer-thick depletion region. The maximum voltage these devices can produce is equal to the semiconductor electronic bandgap. Interestingly, domain walls can give rise to a fundamentally different mechanism for photovoltaic charge separation, which operates over a distance of 1–2 nm and produces voltages that are significantly higher than the bandgap. The authors present in Chapter 10, the charge separation that happens at previously unobserved nanoscale steps of the electrostatic potential that naturally occur at such ferroelectric domain walls. Their nanoscopic size and flexible arrangement using thin film growth engineering solutions and applied external electric fields offer unique possibilities for novel concepts in oxide based photovoltaics.

In: Photovoltaics
Editor: Monique A. Gill, pp. 1-35
ISBN: 978-1-63117-843-6

Chapter 1

NANOSTRUCTURE-BASED PHOTOVOLTAICS

Latha Nataraj*
U. S. Army Research Laboratory, Aberdeen Proving Ground, MD, US

Abstract

Increasing demands on energy has necessitated the development of large-scale, low-cost, renewable energy. Solar energy is highly attractive and extremely promising as the energy reaching the earth from the sun is orders of magnitude larger than today's global energy consumption and is quiet, requires low maintenance, has no restrictions on size, can be located close to the point on consumption, and is a clean, direct and efficient mode of energy conversion. The challenge lies in making solar power economically competitive while working with the materials readily available in abundance along with all their inherent constraints, also using those materials as economically as possible. Novel materials and structures are constantly being investigated for possible application in the area of photovoltaics (PVs). Lately, various nanoparticle-polymer-based solar cells have demonstrated reasonable power conversion efficiencies (PCE), about 9% [1]. Theoretical predictions suggest that higher efficiencies are achievable through nanocrystal-based approaches due to the outstanding optoelectronic properties and multi-exciton generation capabilities of semiconductor nanocrystals (NCs) [2]. Hence, these hybrid solar cells have the potential to provide cost-effective, high-efficiency solutions for PV, although significant challenges need to be overcome.

This chapter discusses several approaches in pursuit of efficient light absorption in materials. Group-IV materials such as silicon (Si) and germanium (Ge) are desirable for optoelectronic applications due to their high abundance, cost-effectiveness, and high compatibility with the existing technology but as they are indirect band gap materials, they inherently have poor optoelectronic properties at room temperature. Approaches such as point defects [3, 4, 5], optically excited Raman lasers [6], quantum confinement in Si NCs [7, 8], utilizing rare earth dopants [9], bonding of III-V semiconductor structures onto Si chips [10], integration of active conjugated-polymers with Si [11], synthesis of tensile-strained semiconductor heterostructures [12] and NCs [13] to modify their band structure and obtain favorable optical properties which result in improved carrier mobilities have been tried to address the challenge.

Band engineered nanoparticles (NPs) and NCs provide improvements in the primary photoconversion process by increasing the energy absorbed from the electromagnetic

* E-mail address: latha.nataraj.ctr@mail.mil

spectrum that may otherwise be lost as heat in bulk semiconductors and thin films. Such NS may be embedded in a suitable matrix for application in PV and could be a potential solution to surpass the ~33% Shockley-Queisser (SQ) limit for PCE, leading us towards viable "third generation" PV devices. This approach also offers great flexibility, allowing us to combine various materials to maximize the range of wavelengths absorbed by the materials which could lead to higher efficiencies for PVs. Also, various nanostructured materials are being experimented for the contact layers. The following sections discuss the synthesis and characterization of the structural and optoelectronic properties of such nanostructures (NSs) and their advantages for high efficiency PV.

Introduction

Nanotechnology offers great promise to revolutionize a vast majority of areas of science and technology as it allows for the manipulation of material properties at the atomic level. Low-dimensional semiconductor structures could provide significant improvement in several applications in the fields of electronics, telecommunications, and energy. NS based PV devices could lead us towards high-efficiency, economical solutions for energy production.

PV devices produce direct current using the light incident upon them. The low dimensions of NSs play a key role in increasing the amount of energy absorbed from the incident light primarily based on two parameters, the material and the size of the NSs. In addition, the band structure of the material may be further influenced by the introduction of strains [13] and doping [14, 15] in these NSs to provide enhanced optoelectronic properties. The key goals of this chapter is to introduce the reader to the basics of semiconductor NSs, approaches for the synthesis of NSs, various techniques used to study their structural and optoelectronic properties, and the fabrication processes for third generation NS-based PV devices.

Semiconductor Material Properties

Crystalline materials have electrons arranged in various energy levels consistent with Pauli's exclusion principle. These energy levels form bands separated by regions of energy where no electrons and wavelike electron orbitals can exist. These forbidden regions of energy caused by interactions between the conduction electrons and the ion cores of the crystal [16] are called "energy gaps" or "band gaps" and strongly influence the electrical and optical properties of materials. Band gaps may be of different widths depending on the properties of the atomic orbitals from which they arise; the band gap is typically inversely proportional to the atomic weight of the material. Sometimes, allowed bands overlap to form a single large band. Electron transfer from one band to the other is possible by means of the processes of carrier generation and recombination.

The upper range of energies of the energy band is called the conduction band. Here, the electrons are allowed to move freely within the crystal lattice. Electrons at these energy levels are mobile charge carriers, responsible for conduction of electric currents in good electrical conductors. The lower range of energy levels where electrons are bound to individual atoms forms the valence band. The bound electrons cannot move freely to be available for conduction.

Electrical and Optical Properties of Semiconductors

Electrical Properties

If allowed energy bands are completely filled or completely empty, no electrons can move in response to an applied electric field and the crystal behaves as an insulator. Partially filled bands give rise to the behavior of conductors (metal). Slightly filled or slightly empty band(s) results in semiconductors which have electrical conductivity values in between that of a conductor and an insulator.

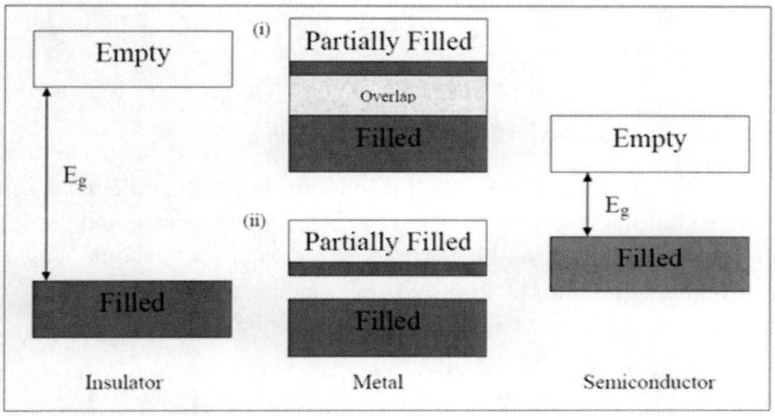

Figure 1. Typical band structure at 0 K for metal, insulator, and semiconductor.

A schematic representation of the band structures of the three types of materials is depicted in Figure 1. The fundamental difference between the band gaps of semiconductors and insulators is that the band gap between the valence band (VB) and conduction band (CB) is much larger in insulators. This prevents electrons being able to get over the larger barrier, therefore resulting in lower, almost negligible electrical conductivity. With thermal energy being a prominent mechanism by which electrons get excited to the conduction band, conductivity of semiconductors is strongly dependent on the temperature of the material.

Optical Properties

The process of optical emission in semiconductors is a result of the transition of an electron from a higher energy level in the CB, across the band gap, into a lower energy level in the VB, and the energy difference being released as a photon, an elementary quantum of light. Therefore, the amount of energy of the emitted photon is dependent on the width of the band gap.

It is possible to stimulate light emission from a semiconductor by means of optical excitation of the material. In doing so, electrons are excited from the VB (E_1) into the CB (E_2) and light emission is stimulated as the excited electrons transition back into the VB to recombine with the hole. The exciting photon energy ($h\upsilon_1$) needs to be at least as much as the band gap energy ($E_2 - E_1$) of the material to be able to excite an electron into the CB. When the electron-hole pairs undergo recombination, the generated photons have energies ($h\upsilon_2$) corresponding to the magnitude of the band gap, as indicated by the schematic in figure 2.

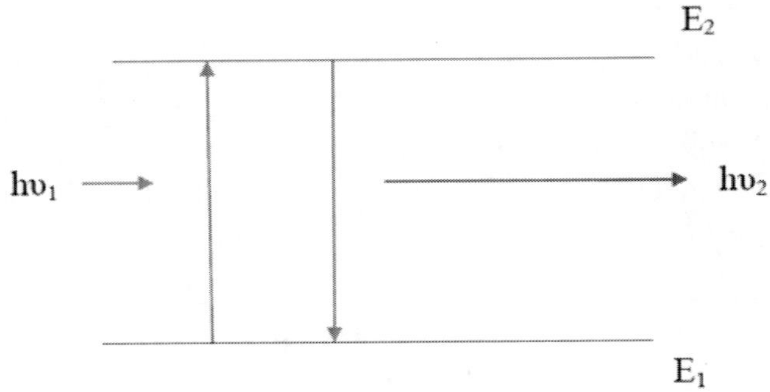

Figure 2. Simplified two energy level model for light emission.

In reality, the energy of the emitted light is slightly lower than the band gap energy. This is attributed to the coulombic attraction between the excited electron and the localized hole it left behind in the VB. The electron-hole pair, referred to as an "exciton" has a lower energy compared to the free unbound electron and hole. Consequently, the energy of the photon released by their recombination is slightly lower than the band gap energy.

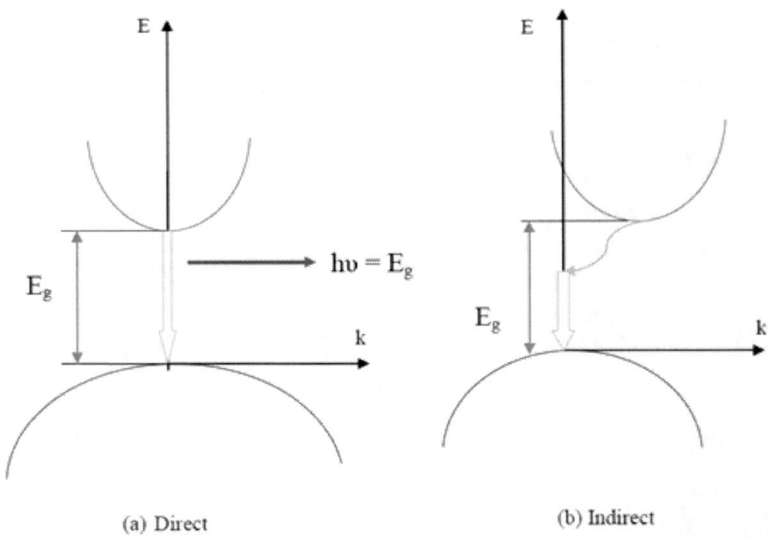

Figure 3. Direct and indirect electron transitions in semiconductors.

Band gaps of semiconductors are classified into direct and indirect types based on the positions of the lowest point in of the CB edge (in green) and the highest point of the VB edge (in red), as shown in figure 3. The bands of energy are distributed along the k-space (reciprocal lattice space) of the crystal in wave-like patterns. If the lowest point of the conduction band occurs at the same k-vector in the Brillouin zone (coordinates in k-space) as the highest point of the valence band as in figure 3a, the band gap is said to be a direct gap; else, it is referred to as an indirect band gap (3b).

The type of band gap has a tremendous influence over the optical properties of a material. In case of direct band gap materials, the transition of the electron for VB to CB is possible by a simple process, where an electron at the top of the VB jumps across the band gap to the bottom of the CB; similarly, the recombination of the excited electron at the bottom of the CB with a hole at the top of the valence band is easy, releasing the excess energy as a photon with the energy of magnitude equal to the band gap. This makes optical absorption and emission highly efficient in direct band gap materials. On the other hand, this would be impossible in an indirect band gap semiconductor as conservation of crystal momentum would be violated. For radiative recombination to occur, the process must also involve the absorption or emission of a phonon (quantum of vibration in the crystal lattice), whose momentum equals the difference between the electron and hole momentum. Interactions among electrons, holes, phonons, and photons are essential for conservation of energy and the crystal momentum (conservation of total k-vector). The phonon requirement results in lower probability of occurrence of radiative recombination in indirect band gap materials. Therefore, optical absorption and emission are far slower and less efficient in indirect band gap materials compared to these processes in direct band gap materials.

The band gap of semiconductors may be engineered in a variety of ways to tune their optical properties such as absorption, emission wavelength, frequency range, and these properties at certain temperatures. Some ways to achieve this are by synthesizing innovative nanoscale structures such as QWs, QDs, and strained NCs to take advantage of quantum effects, synthesizing layered materials with alternating compositions, and creating dislocation loops to eliminate nonradiative recombination pathways. Another means of modifying band gap is by controlling the composition of the individual materials that make up semiconductor alloys such as SiGe, GaAlAs, InGaAs, and InAlAs.

In the following sections we will discuss processes to synthesize NCs, characterization techniques to study their structural, optical, and electrical properties, the fabrication of PV devices using these NSs, and the techniques used to characterize these PV devices.

Nanocrystal Synthesis

Various approaches such as magnetron cosputtering [48, 49, 33, 36], ion implantation [50, 51], isotropic chemical etching into porous Ge [52], thermal decomposition [53], solid-phase epitaxy [54], sol-gel chemistry [55] chemical vapor deposition (CVD) [56, 57], and molecular-beam epitaxy (MBE) [14], and top-down mechanical grinding techniques [30, 13] have been used to fabricate optically-active Ge NCs. In this section, we will discuss briefly each of these processes.

Sputtering sources employ magnetrons, a high-powered vacuum tube generating microwaves by confining charged plasma particles close to the surface of the sputter target using the interaction between a stream of electrons and a magnetic field, as illustrated in its simplest form in figure 4. When the stream of electrons is placed in a magnetic field, the electrons follow helical paths around magnetic field lines. This allows for the plasma to be sustained at a lower pressure and leads to more ionizing collisions with gaseous neutrals near the target surface. Extra ions of the sputtering gas, typically Argon, are created due to these collisions, leading to higher deposition rates.

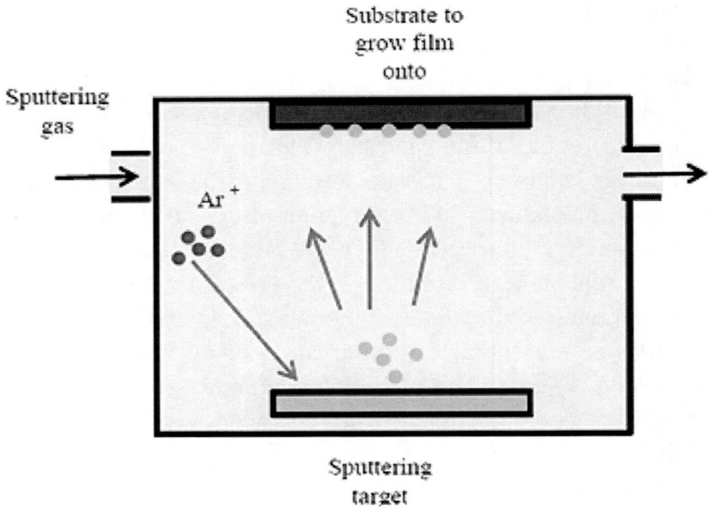

Figure 4. Schematic representation of a Sputtering system.

RF or DC sputtering may be used for the deposition. DC sputtering uses a constant voltage to accelerate the ions to the bombarding velocity required. As the ions strike the surface, the resulting charges move freely about the material to prevent charge buildup. RF sputtering uses a high frequency alternating current to accelerate the ions. Lower frequencies cause the charge to remain localized after the ions strike the surface and over time, the charge builds up, making it impossible to further bombard the surface.

Zacharias et al. use dc-magnetron sputtering in an argon/water atmosphere and a crystalline Si target covered with Ge pieces to prepare Ge NCs in Silica matrix. Maeda et al. [49], Choi et al. [[33], and Takeoka's group synthesize Ge microcrystals and NCs embedded in SiO_2 glassy matrices by RF-magnetron cosputtering.

Ion Implantation is a process by which ions of a material are impacted into a solid using acceleration in an electrical field. The equipment typically consists of an ion source, where ions are produced, an accelerator, where the ions are electrostatically accelerated to a high energy, and a target chamber, where the ions impinge upon the target material to be implanted. Each ion is typically a single atom or molecule, and thus the actual amount of material implanted in the target is the integral over time of the ion current, typically in the order of microamperes. Ion energies are generally in the range of 10 to 500 keV. Energies in the range 1 to 10 keV may be used but provide smaller penetration depth, a few nanometers or less. Lower energies than this result fall under the designation ion beam deposition as the target is hardly affected by the ions. Higher energies of up to 5 MeV may be used, but could cause structural damage to the target.

Kanemitsu et al. have used ion implantation to embed Ge ions in SiO_2 matrix and followed up this process with thermal annealing with temperatures in the range of 300° C to 1000° C thereby being able to achieve a smaller fluctuation in the size of the NCs compared to other methods by them and other researchers [51, 58, 59, 36]. Xiao et al. [54] have used solid phase epitaxy in combination with Ge implantation to obtain Ge-Si nanoclusters, coherently embedded in monocrystalline Si matrix. Kartopu et al. have synthesized Nanocrystalline Ge films by isotropic chemical etching on single-crystalline Ge substrates with 100 and 111 orientations [52]. Monodisperse Ge NCs have been synthesized by Wu et al. [53] by decomposition of a new kind of precursor tetra-*cis*-9-octadecenoic germanium (Ge[CH3*(*CH2*)*7CH=CH*(*CH2*)*8]4, TOG) at high temperature and argon atmosphere. The average grain size can be tuned from 5.2 to 20.4 nm. TOG was first prepared by GeCl4 reacting with oleylamine in a glove box. Ge NCs with narrow size distribution were synthesized by decomposition of TOG at high temperatures. Yang et al. [55] have demonstrated the preparation of Ge microcrystallites by a sol–gel process using 3-trichlorogermanium propanoic acid (Cl3–Ge–C2H4–CO2H) as a Ge source. The Ge content in the starting materials determines the sizes of Ge microcrystallites.

CVD is a popular mechanism to grow NSs. In typical CVD, the substrate is exposed to one or more volatile precursors which chemically react with reacting and/or decompose on the substrate surface to produce the desired deposit. Any volatile by-products produced are removed by gas flow through the reaction chamber. A common classification of CVD methods is based on the operating pressure.

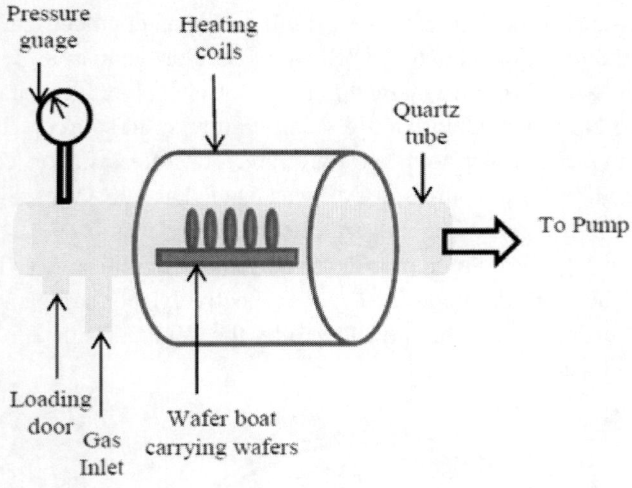

Figure 5. Schematic of a low pressure CVD system.

Palange et al. [56] have prepared Ge 3-D Ge NS on Si using low pressure CVD (LPCVD), such as the one illustrated in figure 5. By changing the deposition parameters they have been able to reproduce the three known modes for such heterostructures and have demonstrated that it is possible to identify confinement effects for samples without cap layers even for structures much larger than the Bohr exciton radius in this material. Ge NCs have been synthesized using ultra-high vacuum CVD (UHVCVD) by Lee et al. [57]. C_2H_4 mediations were used to modify the Stranski–Krastanow growth mode of Ge dots on Si(0 0 1) at 550 °C by ultra-high vacuum chemical vapor deposition. With appropriate C_2H_4-mediation to modify the Si surface, the elongated Ge hut clusters can be transformed to highly uniform Ge domes with a high Ge composition at the core. These C_2H_4-mediated Ge dots, almost bounded by {1 1 3} facets, have an average diameter and height of 55 and 9 nm, respectively.

Molecular Beam Epitaxy is yet another popular method to grow NSs. The growth takes place in high vacuum or ultra-high vacuum (10^{-8} Pa) chamber. The most important aspect of MBE is the low deposition rate, typically less than 3000 nm per hour which allows the epitaxial growth of films. The term "beam" refers to the beam of evaporated atoms which do not interact with each other or vacuum chamber gases until they reach the wafer due to the long mean free paths. The absence of carrier gases in combination with the ultra high vacuum environment provides highest achievable purity levels in the films. In solid-source MBE, ultra-pure elements are heated in separate quasi-Knudsen effusion cells until they begin to slowly sublime. The gaseous elements then condense on the wafer, where they could react with each other. For example, with the elements gallium and arsenic, single-crystal gallium arsenide is formed.

During operation, reflection high energy electron diffraction (RHEED) is commonly used to monitor the growth of the crystal layers. A computer provides precise control over the thickness of each layer through the use of shutters in front of each furnace allowing for fabrication of intricate structures of layers of different materials, leading to the development of structures where the electrons can be confined in space, such as quantum wells or quantum dots. Such layers are now a critical part of many modern semiconductor devices, including solar cells, semiconductor lasers, and light-emitting diodes. More details regarding this technique will be discussed further in the section focusing on characterization techniques.

Sustersic et al. have grown Ge NCs on Si by molecular beam epitaxy using the electron-beam evaporation of Si and the thermal evaporation of Ge. They have also grown the highly advantageous doped NSs. For doping, a high temperature solid source effusion cell was used for boron evaporation. For phosphorus evaporation, a custom effusion cell with a pyrolytic boron nitride baffle and a solid source of GaP was used. The substrates were *n*-type Si _100_ with resistivities of 1–10 _ cm. The substrate temperature was 605 °C during the epitaxial growth of all samples, measured by a pyrometer corrected for emissivity. The nominal growth rates for the Si and Ge were 0.33 and 0.12 Å/s, respectively. For sample A _SGC 520_, a 90 nm undoped Si buffer layer was grown followed by the deposition of 7.5 ML of Ge, forming the undoped Ge QD layer.

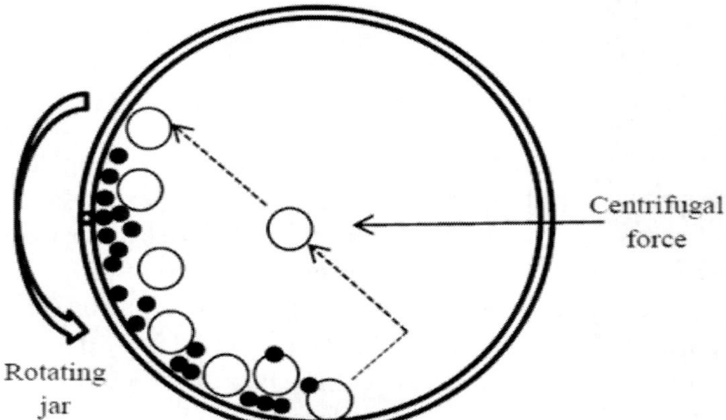

Figure 6. Schematic of a high-energy ball-mill.

Freestanding Ge NCs have been synthesized using a facile mechanical milling process by Giri's team [30] and Nataraj et al. [13, 15]. Mechanical milling may be performed using a simple pestle and mortar set or a high-energy ball mill. During the ball milling process, bulk material placed in the ball mill is subjected to high-energy collision from the balls using a motor to facilitate the rotation of the mixing bowl, as shown in figure 6. In order to minimize contamination of the synthesized nanomaterial, it is common practice to use an extremely hard material such as agate or zirconium oxide for the mixing bowl and grinding media for this process.

This top-down mechanical milling process used to synthesize undoped and doped NCs inherently provides less control over the size uniformity. A wide size distribution ranging from a few nm to hundreds of nm has been observed. These NS may be used as is in applications where the size uniformity is not critical, such as in PV. For applications where size uniformity is essential, one could employ advanced filtering mechanisms such as a nanosieve[60] to filter out NS of a certain size.

Nanomaterials Characterization Techniques

NSs synthesized using the various fabrication processes discussed in the previous section have promising applications in novel, third-generation PV devices. It is therefore essential to confirm the structural integrity of these NSs and understand the fundamental physical properties. Various characterization techniques are used to gain insight into the various properties of these structures, namely structural properties such as defect densities and crystal quality, mechanical properties such as lattice strain, and optical properties such as photo-emission and absorption characteristics.

Structural properties may be investigated using Atomic Force Microscopy (AFM), Scanning Electron Microscopy (SEM), Selected Area Electron Diffraction (SAED), Surface Enhanced Raman Spectroscopy (SERS), Transmission Electron Microscopy (TEM), and X-ray Diffraction (XRD). Optical properties have been studied using photo emission and absorption measurements and their dependence on temperature and excitation intensity using techniques such as Optical absorption spectroscopy, Photoluminescence (PL) spectroscopy, Temperature-dependent PL spectroscopy, Excitation intensity dependent PL spectroscopy, Time-resolved PL spectroscopy, and Photocurrent spectroscopy.

AFM is a very high-resolution type of scanning probe microscopy. Using the mechanical movement of a physical probe in a raster scan of the specimen, and by recording the interaction between the probe and the surface of the specimen as a function of the probe position, it generates images of surface of the sample.

As illustrated in figure 7, the probe consists of a cantilever, typically Si or silicon nitride that is about 100 μm long, 20 μm wide and 0.1 μm thick, with a tip radius of curvature on the order of nanometers attached to a sharp tip used to scan the specimen surface. Based on the design, the cantilever moves over the sample or the sample is translated under the cantilever by the piezoelectric scanner.

When the tip is brought to the proximity of the sample surface, forces between the tip and the sample causes a deflection of the cantilever in accordance with Hooke's law. The sensing of the cantilever motion is commonly done by detecting the light reflected from the cantilever into a two-segment or four-segment position sensitive photodiode as shown in figure 7. The reflected light impinges upon different segments of the photodiode depending on the cantilever motion. The horizontal motion is recorded using x = [(A+B) – (C+D)] and the vertical motion is given by [(A+C)-(B+D)] [151], where A, B, C, and D are different regions of the photodiode.

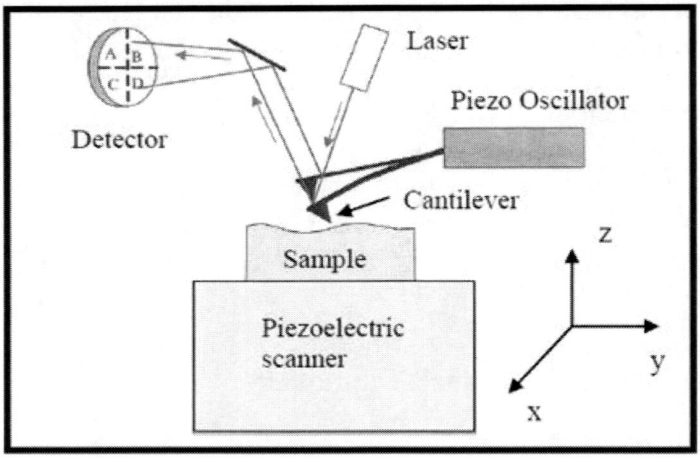

Figure 7. Schematic of AFM set-up.

AFM can operate in different modes. In the contact mode, the probe tip dragged across the surface results in a topographical map of the sample surface. In this mode, the dragging could result in damage to the sample and the probe as well as artifacts in the data. In addition, capillary action causes a meniscus to form from condensed water vapor and other contaminants that generally on the sample surface under normal ambient air conditions and surface tension causes the tip to get pulled towards the sample surface. This attractive force, in addition ot adhesive forces due to the electrostatic charge accumulated in the tip and on the surface causes the overall force on the sample to increase. This increased force in addition to the lateral shear force due to the scanning can lead to distorted measurement data and damaged sample.

The AFM may also be operated in a non-contact mode to avoid tip or sample degradation effects. It works by sensing the weak Van der Waals attractive forces between the tip and the sample. In this mode, the cantilever is oscillated at a frequency slightly above its resonance frequency which is given by

$$f_0 = \frac{1}{2\pi}\sqrt{\frac{k}{m}}, \quad (1)$$

where k is the spring constant and m is the mass of the cantilever. The amplitude of oscillation is generally in the order of a few nanometers. The Van der Waal force acts to decrease the resonance frequency of the cantilever which acts along with the feedback system to maintain the specified amplitude or frequency by adjusting the average distance between the sample and the tip of the cantilever. The distance at each data point measured allows the scanning software to construct a topographic image of the sample surface.

Another mode of operation is called the tapping mode, where a small piezoelectric element mounted in the AFM tip holder drives the cantilever to oscillate up and down at or close to its resonance frequency with the amplitude in the order of a few hundred nanometers. Due to the interaction of forces such as Van der Waals force, dipole-dipole interaction, or electrostatic forces acting on the cantilever due to the proximity of the tip to the sample surface, the amplitude of oscillation decreases as the tip gets closer to the sample. An electronic servo and a piezoelectric actuator provide the feedback to control the height of the cantilever above the sample surface by adjusting the height to maintain the cantilever oscillation amplitude and force as the cantilever is scanned over the sample. An image is generated using the changes in force of the oscillating contacts as the tip scans the sample surface. Tapping mode is an improvement over contact AFM as the cantilever does not need to be dragged over the sample surface, therefore avoiding surface damage. This mode is effective for soft, adhesive or fragile samples that pose challenge to imaging or can be easily damaged by other imaging techniques.

Electron Microscopy

An electron microscope uses an electron beam to raster scan the sample with a high energy beam (e-beam) of electrons, resulting in a magnified image of the sample. The use of electrons provides two major advantages over optical microscopy:

E-beam wavelengths being much smaller than the photon wavelengths, the electron microscope provides much higher resolution compared to optical microscope and higher magnification. If υ is the electron velocity and V, the accelerating voltage, the e-beam wavelength, λ_e, is given by [61]

$$\lambda_e = \frac{h}{m\upsilon} = \frac{h}{\sqrt{2qmV}} = \frac{1.22}{\sqrt{V}} = 0.012 \text{ nm} \quad (2)$$

for V=10,000 V << 400 nm – 700 nm for visible light

Higher depth of field

Scanning electron microscope (SEM) and Transmission electron microscope (TEM) are among the principal types of electron microscopes in use.

Scanning Electron Microscopy (SEM)

As an SEM raster scans the sample surface with a focused e-beam, various signals are generated as a result of the interactions between the electron beam and atoms at or near the surface of the sample such as secondary electrons, backscattered electrons (BSE), X-rays, emitted light (cathodoluminescence), and e-beam induced current due to absorbed electrons. Commonly, SEMs consist of secondary electron detectors, but could also have detectors for other possible signals. The volume of the generated signals is a function of the e-beam energy and the atomic number of the sample. Most electrons penetrate deep into the sample and are absorbed in case of low-Z samples. In high-Z samples, a large portion of the incident electrons is backscattered. One of the empirical expressions for the electron penetration depth, also termed the *electron range, Re* is

$$R_e = \frac{4.28 \text{ x } 10-6 \text{ E}1.75}{\rho} \text{ cm} \quad (3)$$

where ρ (gm/cm3) is the sample density and E (keV) is the energy [62]. Equation 3 is fairly accurate for 20 < E < 200 keV, but slightly underestimates the range.

Secondary electrons are collected by a detector, which is not always possible in conventional optical microscopes. Light reflected from the sample, after passing through a lens for amplification, forms an image by mapping and transforming the information from sample space to CRT space resulting in the displayed image. In secondary electron imaging (SEI), the standard detection mode, the SEM produces very high resolution images, revealing details as small as 1 nm - 5 nm in size.

SEM micrographs have a large depth of field due to the very narrow electron beam. This results in the characteristic 3-D appearance in the image of the topography of the sample. SEM offers a wide range of magnifications, from about 10 times (like a powerful hand-lens) to more than 500,000 times (about 250 times the magnification limit of the best conventional optical microscopes). The SEM contrast is a function of several factors. While a flat uniform sample exhibits no contrast, a sample with non-uniform topography acts as a significant SEM contrast enhancement feature. The variation of the contrast C with the angle if the incident electron beam is given by the following expression [18]:

$$\text{C} = \tan(\theta)\, d\theta \quad (4)$$

Thus steep surfaces and edges tend to be brighter than flat surfaces, which results in images with a well-defined, three-dimensional appearance. Using this technique, image resolution less than 0.5 nm is possible.

The resolution of e-beam measurements in semiconductor samples is not always good although the beam diameter is 0.1-0.4 nm, due to elastic scattering caused by the interactions with the nuclei, generally with the case of higher Z materials and lower e-beam energies. In addition, they also undergo inelastic scattering from valence and core electrons. These scattering mechanisms broaden the originally collimated e-beam, resulting in decreased resolution. However, SEM is highly useful for the study of the surface of semiconductor devices, especially during fault analysis and cross-sectional analysis. The surface is usually coated with a conductive material like gold to reduce/eliminate the surface charging or by reducing the e-beam energy to ~1 keV to ensure that the primary electrons is roughly equal to the secondary and back-scattered electrons, and increasing the beam brightness to improve the signal-to-noise ratio.

Transmission Electron Microscopy (TEM)

Transmission electron microscopy (TEM) is an electron microscopy technique which uses a beam of electrons transmitted through an ultra-thin sample, and generates an image using the interaction between the electrons and the sample; the image is magnified and focused onto an imaging device or to be sensed by a detector such as a CCD camera.

TEMs provide significantly higher resolution than conventional optical microscopes, due to the small de Broglie wavelength of electrons. The resolution is given by the equation,

$$s = 0.61\lambda / NA \tag{5}$$

where s is the resolution, λ is the wavelength of incident beam, and NA is the numerical aperture. In optical microscopes, NA = 1, λ = 500 nm, and therefore, s = 300 nm. In case of TEM, NA = 0.1, λ = 0.004 nm for a voltage of 100 kV from equation 2, and therefore, s = 0.25 nm. Thus TEM provides orders of magnitude high magnification compared to the optical microscope, enabling the examination of fine detail small as a single column of atoms.

At smaller magnifications, absorption of electrons in the material variable with thickness and composition of the material determines the contrast in the TEM image. At higher magnifications, complex wave interactions modulate the intensity of the image. Using ultra-thin samples of thicknesses ranging over a few tens to a few hundred microns overcomes the limited depth resolution of TEM as it prevents the e-beam from spreading. Various modes of TEM allow for the observation of modulations in several properties such as chemical composition, crystal orientation, and electronic structure of materials. TEM has increasingly become an important tool in a range of scientific fields, including cancer research, virology, materials science, environmental pollution and semiconductor research.

High resolution TEM (HRTEM) is yet another mode of TEM that enables the imaging of the crystalline structure of a sample at an atomic scale. Its high resolution, of the order of angstroms makes invaluable for the study of nanoscale properties of crystalline materials like semiconductors and metals as individual atoms and crystalline defects can be imaged. It is possible to obtain a 3-D map of the material by combining several views of the crystal, taken from different angles, a technique called electron crystallography. HRTEM image formation relies on phase-contrast which is not necessarily intuitively interpretable due to strong aberrations such as astigmatism of the imaging lenses in the microscope, which could be estimated from the Fourier transform of the HRTEM image.

Selected area electron diffraction (SAED), can be performed inside TEM. The small de Broglie wavelength of high-energy e-beam and comparable to the atomic spacing in the sample, the atoms act as a diffraction grating and scatter some of the electrons at particular angles as determined by the sample crystal structure, while others are transmitted without deflection, resulting in a TEM image that appears as a series of spots, each spot corresponding to a satisfied diffraction condition of the sample's crystal structure.

The user can easily choose from which part of the specimen to obtain the diffraction pattern. A selected area aperture, a thin strip of metal that will block the beam, is located below the sample holder on the TEM column can be inserted into the beam path. With several different sized holes, it can be moved to select the area of interest, blocking all of the e-beam except for the small fraction passing through one of the holes. Similar to x-ray diffraction (XRD), SAED provides information about crystal structures and crystal defects. In addition, it also has the advantage of being able to examine areas as small as several hundred nanometers in size, while XRD typically samples areas of the order of several centimeters in size.

Reflective High Energy Electron Diffraction (RHEED)

Reflection high-energy electron diffraction (RHEED) is a characterizing technique to study the surface of crystalline materials based on diffraction of high-energy electrons. The basic setup of a RHEED system is as illustrated in figure 8, with an electron gun, a photo luminescent detector screen and the sample. The e-beam is generated by the electron gun and strikes the surface of the sample at a very low angle of incidence, typically less than 5°. Regular patterns are formed on the detector as incident electrons diffract from the atoms at the surface of the sample and the forward scattered electrons interfere constructively at specific angles. The diffraction pattern at the detector is a function of the sample surface as the electron interference is influenced by the position of atoms on the sample surface. Thus, the RHEED provides information on the surface crystal structure, surface orientation, and surface roughness. It is commonly used during epitaxial growth for continuous monitoring during MBE.

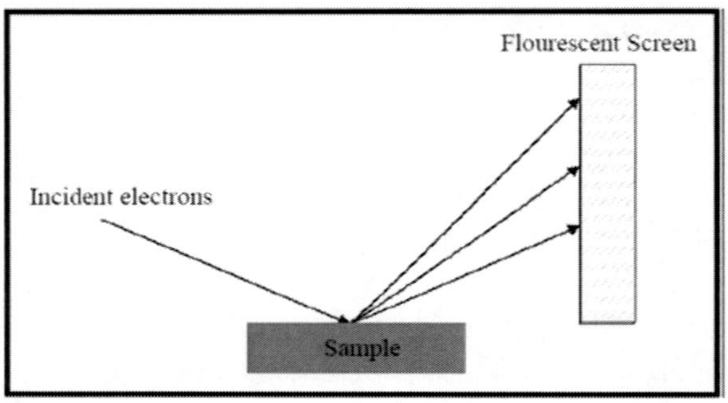

Figure 8. Schematic of RHEED diffractometer.

X-Ray Diffraction

X-ray diffraction is a non-destructive technique which gives structural information over the entire sample using X-rays based on the elastic scattering of X-rays from crystalline structures. It is commonly to determine structural crystal defects.

As depicted in the arrangement in figure 9, a perfect crystal is arranged to diffract monochromatic X-rays of wavelength λ from the lattice planes separated by a distance d. X-rays impinge upon the sample with an angle of incidence α. The primary beam is absorbed by or transferred through the sample and the diffracted beam is recorded on film. The angle of emergence of the diffracted beam is twice the Bragg angle, θ_B, and is given by the following equation [61]:

$$\theta_B = \arcsin(\lambda/2d) \qquad (6)$$

The diffracted rays are detected using a solid state sensor or recorded on a high-resolution photographic film such that the detector or film is perpendicular to the diffracted rays for highest resolution and is held close to the sample without intercepting the incident beam. Structural defects and variations in the lattice plane or orientation defies equation 6, resulting in a difference between the intensities of the diffracted rays from the two regions.

XRD may also be used to determine elastic strain in the sample, which is defined by the equation below [61]:

$$S = \frac{d_{unstrained} - d_{strained}}{d_{unstrained}} \quad (7)$$

where $d_{unstrained}$ and $d_{strained}$ are determined by equation 6.

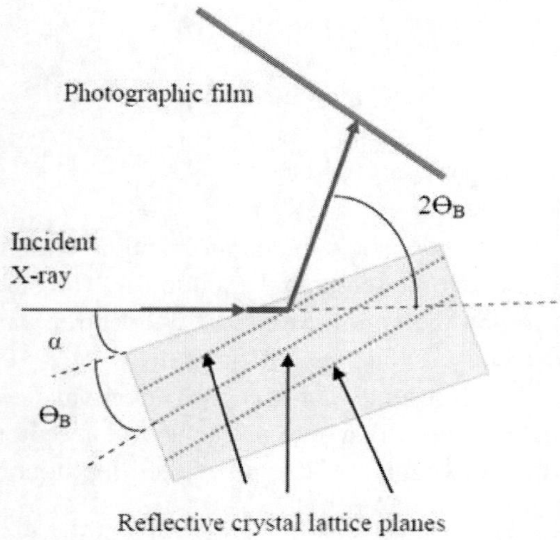

Figure 9. Schematic of XRD system.

Therefore, XRD is useful determining a variety of characteristics of the material such as surface quality, lattice spacing, strain, and defect characteristics.

Photoluminescence Spectroscopy

Photoluminescence (PL) spectroscopy is a nonintrusive method of probing the electronic structure of materials by photoexcitation of the sample which absorbs the incoming light energy. One means of dissipation of this excess energy by the sample is by light emission or luminescence. The intensity and spectral content of this photoluminescence is a direct measure of various important material properties.

Photo-excitation causes electrons within the material to move into higher allowed states. When these electrons relax to their equilibrium states, the excess energy is released and may include radiative or non-radiative processes. The intensity of the emitted light is related to the relative contribution from the radiative processes and the energy corresponds to the difference in energy levels involved in the downward transition and is determined by the recombination process where five types of transitions may be observed [61] as shown in figure 10.

- Band-to-band recombination is predominant at room temperature. It is rare at low temperatures for materials with small effective mass due to large electron orbital radii. It is indicated by the transition a in figure 10.
- Excitonic recombination is commonly observed (b). The e-h pair is free to move through the crystal and as the electron and hole always move together,

photoconductivity and photocurrent are not observed through this process, but light is emitted when they eventually recombine. In pure direct band gap materials,

$$h\upsilon = Eg\text{-}Ex \tag{8}$$

where *Eg* is the band-gap energy and *Ex* is the excitonic binding energy. Indirect bandgap materials require a phonon for light emission for conservation of momentum, and the relationship changes to include the phonon as follows:

$$h\upsilon = Eg\text{-}Ex\text{-}Ep \tag{9}$$

where Ep is the energy of the phonon generated.

- In case of doped materials, a free hole can combine with a neutral donor to form a positively charged excitonic ion/bound exciton (c). The electron of the donor moves in a wide orbit around the donor. Similarly, an electron can combine with a neutral acceptor to form a bound exciton as shown in (d).
- (e) illustrates an the case of donor-acceptor recombination where an electron in a neutral donor recombines with a hole in the neutral acceptor. The emission energy is determined by the coulombic attraction between the donors and the acceptors, and the energy is defined as follows [61]:

$$h\upsilon = E_g\text{-}(E_A + E_D) + \frac{q^2}{K_s\varepsilon_0 r} \tag{10}$$

where EA and ED are the acceptor and donor energy levels respectively and if this quantity is too low, the photon might get reabsorbed into the sample and with energy higher than the band gap energy. Photons might also get reabsorbed if the incident angle is within the critical angle. This leads to the determination of an internal PL efficiency for the material, given by [61]

$$\eta_{int} = \int_0^d \frac{\Delta n}{\tau_{rad}} \exp(-\beta x)\, dx \approx \int_0^d \frac{\Delta n}{\tau_{rad}}\, dx \tag{11}$$

where d is the sample thickness, Δn is the excess minority carrier density, $\tau_{rad}$is the radiative recombination lifetime, and $\beta$is the absorption coefficient of the generated light. With the volume analyzed by PL measurements being dependent on the absorption depth of the excitation source and the diffusion length of the minority carriers, the absorption of light in a medium is specified by the Beer-Lambert law which states that the transmissivity (T) of light through a material depends logarithmically on the product of its absorption coefficient (α) and the distance (l) through which it travels inside the material:

$$T = \frac{I}{I_0} = e^{-\alpha l} \tag{12}$$

where I and I0 are the intensities of the transmitted light and the incident light respectively. The absorbance may be expressed in terms of transmissivity as

$$A = \ln\left(\frac{1}{T}\right) = \ln[e^{\alpha l}] = \alpha l \quad (13)$$

The absorption coefficient, $\alpha = 1/y$, where y, the penetration depth, is the distance through which the incident light travels in the material, after which the flux density drops off by a factor of $1/e \approx 1/3$ its initial value is given by

$$\alpha = \frac{2\omega n}{c} = \frac{4\pi n}{\lambda} \quad (14)$$

where ω is the angular frequency and λ is the wavelength of the incident light and n is the refractive index of the material, and c is the speed of light in vacuum. The absorption depth, dependent on the wavelength of incident light and the refractive index of the material is generally in the order of microns. Energy of excitation can be increased to reduce the absorption depth, useful when working materials like SOI and thin films.

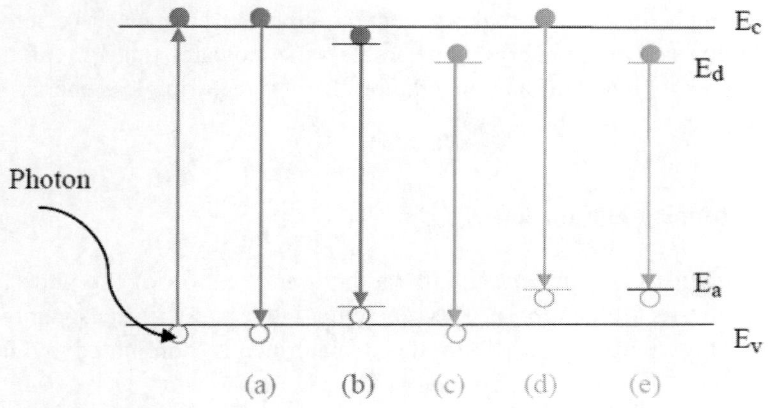

Figure 10. Radiative Transitions.

Photoluminescence is particularly suited for detecting the band gap energy as wells as shallow level impurities and sometimes deep level impurities, if they exhibit radiative recombination. The band gap of a material is approximately the energy of emission from the material. In addition, the energy and intensity of radiative transitions that involve localized defect states as reflected in the PL spectra help identify the defect states and their concentration respectively. Even in the presence of different types of impurities, as long as they produce radiative recombination processes, PL spectra help identification of impurities, but they are not so useful in the measurement of the impurity density.

Variations of PL such as temperature-dependent-PL and excitation-intensity dependent PL are highly useful for understanding the underlying physics in material properties and help to determine whether radiative or non-radiative mechanisms are dominant in the recombination process. Low temperature measurements help filter out thermally activated non-radiative recombination processes and thermal peak broadening and provide comprehensive spectroscopic information by eliminating ~kT/2 influence to the width of the emission line due to distribution of the thermally excited carriers.

Temperature Dependent PL

PL intensity variation with temperature is given by a Boltzmann distribution given by [63]

$$I(T) = \frac{I(0)}{1 + C\exp(\frac{-E_A}{kT})} \tag{15}$$

where I(0) is the intensity at 0 K, I(T) at temperature T, C is a temperature dependent constant, E_A is the activation energy, and k is the Boltzmann constant. Therefore,

$$E_A = kT\,(\ln[\,I(T)] - \ln[I(0)]). \tag{16}$$

The slope of the higher temperature region of the plot of the natural log of the peak intensity against 1/T may be used to calculate the activation energy.

The activation energy is the highest temperature at which PL becomes noticeable. In cases where energy is removed from the system (cooling) we have negative activation energy. In this case, higher activation energy is desirable as it indicates that PL can be observed at higher temperatures, closer to room temperature, thereby requiring less energy to be removed from the system.

Excitation Intensity Dependent PL

Excitation intensity dependent PL allows the identification of the underlying physical mechanisms of the recombination process in semiconductors with reasonably good crystal quality, as the low-temperature PL in these materials is dominated by nearband-edge emission [64].

Table 1. Summary of transitions during near-band-edge PL

Code	Transition
A	Free-exciton recombination (FE)
B, C	Radiative recombination of donor- and acceptor- bound excitons (D^0X, A^0X)
D	Donor-acceptor pair recombination (D^0A^0)
E	Radiative recombination of free electron with neutral acceptor (eA^0)
F	Radiative recombination of free hole with neutral donor (hD^0)
G, H	Non-radiative recombination of free electrons and holes with ionized donors and
I	Excitation of electrons and holes by above-band-gap laser light ($h\upsilon > E_g$)
J	Resonant formation of free excitons by laser light ($h\upsilon = E_g$)
K, L	Excitation of electrons from neutral donors or ionized acceptors by laser energy

Various groups [65, 66, 67, 68] have shown that the variation of PL intensity (I) with excitation laser power (L) is defined by a power law distribution of the form, $I \alpha L_k$, where exponent k takes into account decay mechanisms through radiative and non-radiative recombination processes.

The relaxation pathways of the photo-excited carriers represented figure 12 are summarized in Table 1. Theoretical models indicate that k is in the range $1 < k < 2$ for exciton-like transitions and $k < 1$ for free-to-bound or donor-acceptor pair transitions, and are in good agreement with experimental results [64] [69]. The power index k is

$$k = \frac{d \log I}{d \log L} \tag{17}$$

Therefore, k is apparent slope of the log scale plot of I as a function of L. From [64] and [69], the rate equation for the electron population in the CB is

$$\frac{d n}{d t} = -an^2 - bn + cL \tag{18}$$

where a, b, and c are proportional factors depending on material properties. The first term corresponds to emission by the formation of free excitons while the second and third term account for other radiative and non-radiative recombination processes. At steady state, = 0, and the solution to equation 18 is

$$n = A\,(-1 + \sqrt{1 + BL}) \tag{19}$$

where $A \equiv b/2a$ and $B = 4ac/b^2$. Also, at steady state, the intensity of the free exciton emission, I_{FE} α n^2. Therefore, for the free exciton component of the emission, k may be evaluated from eq. 18 as follows:

$$k = \frac{d \log I_{FE}}{d \log L} = 2\frac{d \log n}{d \log L} = 2\frac{d\,[\log A + \log(-1 + \sqrt{1+BL})]}{d \log L} \tag{20}$$

Therefore,

$$k = 2\,\frac{d \log(-1 + \sqrt{1+BL})}{d \log L} \tag{21}$$

as log A is not a function of log L. For x= BL, we have d log x = d log L as log B does not depend on log L. Therefore,

$$k = 2\,\frac{d \log(-1 + \sqrt{1+x})}{d \log x} = \frac{2x}{2\sqrt{1+x}(-1 + \sqrt{1+x})} \tag{22}$$

The value of x, and therefore k, is dependent upon excitation power, a, b, and c based on material properties. As b includes radiative and nonradiative recombination processes, k is sensitive to radiative and competitive non-radiative probabilities and the excitation laser power. k is plotted for different values of x, in accordance with equation 22. The value of k has been shown to be in the range $1 < k < 2$, for different values of x which depends on L and material properties reflected by parameters a, b, and c.

At a very low excitation power, a small fraction of free electron-hole pairs form excitons and most photoexcited carriers recombine through defect states. Hence, the an^2 term becomes negligible in equation 18, resulting in a linear dependence of I on L. More excitons are formed from free carriers with higher excitation power, minimizing the terms which are linear in n and I varies as $L^{0.5}$. When neither free exciton recombination nor donor- acceptor recombination dominates, the linear and quadratic terms of equation 18 are equal in magnitude and the log-log plot of I vs.L would be a curved plot. From eq. 19, we have

$$x = BL = [\frac{4ac}{b^2}] L \quad (23)$$

From eq. 18, processes represented by a and b compete with each other, meaning probabilities of radiative and non-radiative recombination processes compete with the free exciton radiative recombination. The ratio of radiative to non-radiative processes varies with the crysal lattice; Reduced lattice perfection leads to increased probability of non-radiative processes and dopants increase the probability of radiative processes due to the recombination of free carriers. Increased probability of non-radiative processes is reflected by a higher b, and therefore a lower a, which leads to a reduction of x which in turn results in a higher k. Therefore, a higher k at low pump powers indicates a greater probability of non-radiative recombination processes [70]. A deviation in the high excitation power region indicates thermal effects [70, 71] due to laser-induced heating. Therefore, excitation intensity dependent PL is a powerful means of determining the probabilities of radiative to non-radiative recombination mechanisms in materials.

Time-Resolved PL

Time-resolved photoluminescence is used to gain understanding recombination mechanisms in materials. Direct recombination happens spontaneously, meaning the recombination of probability is constant in time. This leads to an exponential solution for the decay of the excess carriers. This decay rate of carriers at a given time t, is proportional to the number of electrons n, and holes p, at time t, with a constant of proportionality for recombination αr. The difference between the net thermal generation rate and the rate of recombination is the net rate of change of electrons in the conduction band. Thus,

$$\frac{dn(t)}{dt} = \alpha r n i\ 2 - \alpha r\ n(t)\ p(t), \quad (24)$$

where n_i is the intrinsic carrier concentration. At t = 0, the initial generated electron concentration Δn is the same as Δp, the generated hole concentration. As they recombine forming pairs, the instantaneous values of the electron and hole concentrations, δn(t) and δp(t) respectively, are also equal.

Therefore, the total concentration in equation 18 in terms of excess carrier concentrations δn(t) = δp(t) and equilibrium concentrations n_0 and p_0 as follows:

$$\frac{d\delta n(t)}{dt} = \alpha_r n_i^2 - \alpha_r[n_0+\delta n(t)][p_0+\delta\ p(t)] = -\ \alpha_r[(n_0 + p_0)\ \delta n(t)+ \delta^2\ n(t)\ ], \quad (25)$$

as $n_0p_0 = ni^2$.

For low-level injection, as δ n(t) << n0, the δ^2 n(t) term is negligible. When evaluating doped materials, for p-type material, p0 >>n0.

Therefore, equation 15 becomes

$$\frac{d\delta n(t)}{dt} = - \alpha_r p_0 \delta n(t) \tag{26}$$

and has the solution

$$\delta n(t)= \Delta n e^{-\alpha_r p_0 t} = \Delta n e^{\frac{-t}{\tau_n}} \tag{27}$$

where ηn = $-1/\alpha_r p_0$ is called the recombination lifetime of the electrons. The lifetime is calculated in terms of the minority carriers as there is a much higher percentage change in the minority carrier concentration compared to that of the majority carrier concentration, and is referred to as the minority carrier lifetime. Similarly, we have ηp as the minority carrier lifetime in an n–type material.

PL Experimental Set-Ups

The following sections provide brief descriptions of experimental details for the basic PL and variations of this technique. Excitation may be achieved using a laser, with energy $h\upsilon > E_g$ generating electron-hole (e-h) pairs which recombine resulting in radiative recombination and non-radiative bulk or surface recombination, depending on a variety of factors discussed in earlier sections.

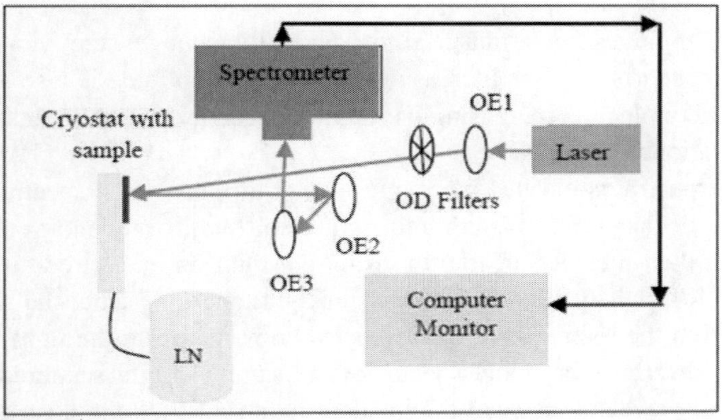

Figure 11. A typical PL set up; O1, O2, and O3 are optical elements lenses/ mirrors.

A typical PL set-up is shown in figure 11. A beam from the excitation laser source is directed and focused onto the sample using optical components. The emitted light is directed into a spectrometer, filtering out the laser reflections and the ambient light which is processed using a computer to display the emission spectrum.

For temperature-dependent PL, the sample mounted on a cryostat is cooled using liquid Helium or liquid Nitrogen, as indicated in figure 11 (LN2). In addition, this figure illustrates a rotary filter system consisting of filters of varying optical densities (OD) in the path of the incident beam for excitation intensity dependent PL, to observe the evolution of the emission spectra with the intensity of the incident light.

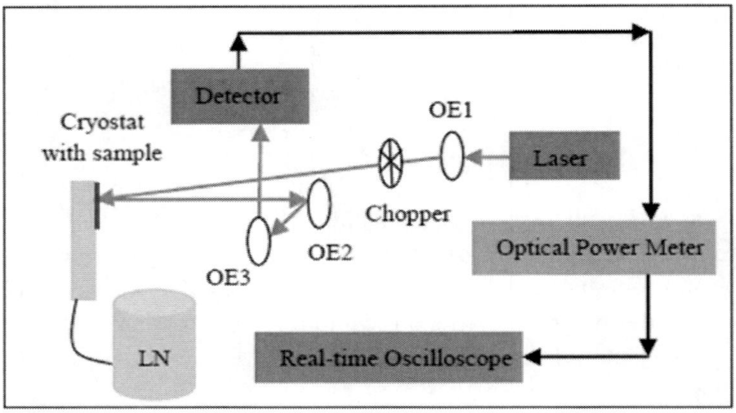

Figure 12. Time-resolved PL set up; O1, O2 and O3 represent optical elements.

Time-resolved PL provides information on the radiative recombination lifetime of the sample. A high speed chopper is used in the path of the incident beam and the emitted light is collected by a detector hooked up to a real-time oscilloscope, as shown in figure 12.

Raman Spectroscopy

When light impinges on a material, it generally changes the vibrational energy of molecules due to absorption. The increasing vibrational amplitude is also sufficient to excite the rotations of the molecule, increasing its rotational frequency, as the rotational energies are smaller than vibrational energies.

Vibrational spectra could be measured using infrared (IR) spectroscopy or Raman Spectroscopy. In IR spectroscopy, infrared light of different frequencies are passed through the sample and the intensity of the transmitted light is measured at each frequency. Transmittance = I_s/I_r, where I_s is the intensity of light emerging from the sample and $I_r$is the light emerging from the reference. Raman spectroscopy measures the light scattered from the sample and provides the vibrational spectra of the sample. Light scattered from the sample surface consists of wavelengths υ_0 of the incident light (elastic/Raleigh scattering) and a very small fraction (of the order of 1/1000), also different wavelengths $\upsilon_i$at very low intensities representing interactions of the incident light with the material. The interaction of the incoming light with optical phonons causes Raman scattering and the interaction with acoustic phonons results in Brillouin scattering. $\Delta E = |\upsilon_0 - \upsilon_i|$ corresponds to the energies absorbed by the sample, which is a vibrational frequency in the infrared range.

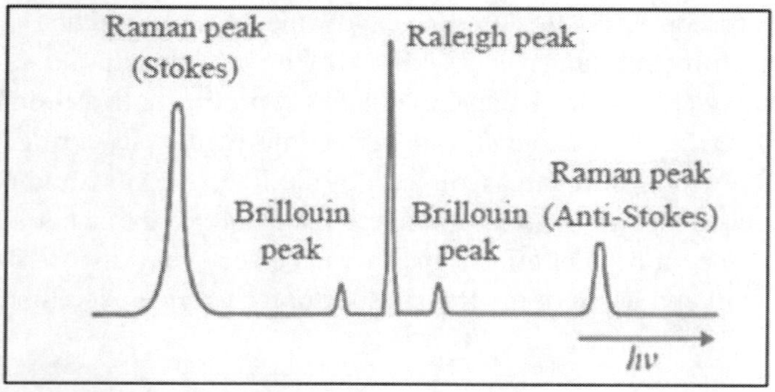

Figure 13. Energy distribution of scattered light.

As illustrated in figure 13, optical phonons have higher energies than acoustic phonons. This results in the optical phonons giving larger photon energy shifts. When an incident photon interacts with the sample and imparts energy to the lattice in the form of a phonon, a low-energy scattered photon ($\upsilon_i < \upsilon_0$) is generated, resulting in a down-shift in energy. This is called Stokes-shifted scattering. When the photon absorbs a phonon from the material and emerges with higher energy ($\upsilon_i > \upsilon_0$), it results in Anti-Stokes-shifted scattering. This is a much weaker mode than the Stokes scattering. The tallest peak in the center in figure 14 corresponds to Raleigh scattering, which is of the same energy as the incident laser (υ_0).

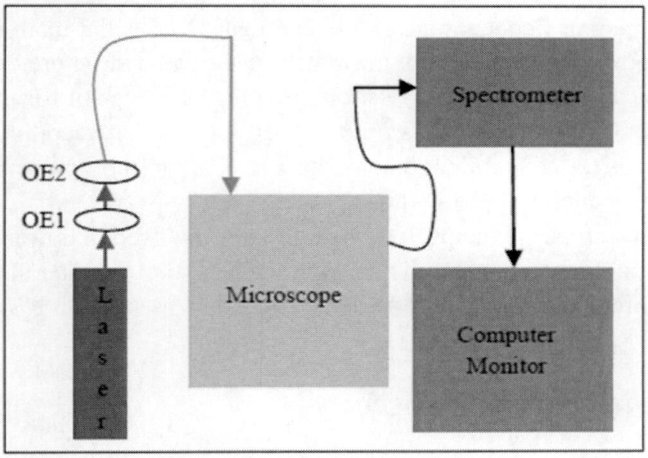

Figure 14. Raman Spectroscopy experimental set-up. O1 & O2 are optical elements.

The experimental set-up for Raman spectroscopy consists of a low power laser, (< 5mW to reduce sample heating) in the visible, near infrared, or near ultraviolet range, that illuminates the sample through a microscope. Scattered light from the illuminated spot is collected through a monochromator to filter Rayleigh scattering, while the rest is dispersed onto a detector. Detection is challenging due to a weak signal against an intense background and is mitigated by observing the Raman signal at right angles to the excitation beam. The detector interfaces with a computer and displays the Raman spectra from the sample. A typical Raman spectroscopy setup is as shown in figure 14.

Raman spectroscopy may be used to study the crystal structure as different crystal orientations give different Raman shifts. By varying wavelengths, it is possible to reach different absorption depths in the sample to obtain the profile up to a certain depth. It helps to monitor implant damage as structural imperfections cause scattering by forbidden TO phonons. Also, the width, shift and asymmetry of the Raman peak could provide information regarding extremely small grain sizes or amorphous nature of the material, stress or strain in the sample. Phonon confinement effects and thermal effects may also be determined from the evolution of the shift and width of the Raman spectra for varying excitation power.

Photocurrent Spectroscopy

Photocurrent spectroscopy is based on the principle of electrical detection of optical signals. By measuring the lateral photocurrent between the two contacts on the same side of the sample that results from the absorbed optical excitation energy, it is possible to obtain information regarding absorption in the material. The current is proportional to the absorption coefficient of the material [73, 74] and for indirect-bandgap semiconductors such Si and Ge, the absorption and incident energy may be expressed with the following relationship:

$$\alpha h\upsilon = \beta\,(E_g + E_{LO} - h\upsilon)^2, \tag{28}$$

where α is the absorption coefficient, $h\upsilon$ is the incident photon energy (where h is Planck's constant, υ is the photon frequency), E_g is the band gap, E_{LO} is the main longitudinal-optical phonon energy corresponding to the k-mismatch in the band structure, and β is an energy-independent physical parameter. This relation provides the linear fit for the absorption band-edge data, and the slope of the fit gives β [75]. Therefore, the intercept of the linear fit on the abscissa corresponds to the incoming photon energy equal to the sum of the bandgap energy and the main phonon energy ($E_{photon} = h\upsilon = E_g + E_{LO}$).

Photocurrent spectroscopy helps to probe the states involved in upward optical transitions and is a powerful tool to understand the absorption properties of materials, which can further be used to make inferences regarding other properties such as strain in NSs.

Optical Absorption Spectroscopy

Absorption spectroscopy is yet another spectroscopic technique to measure optical absorption as a function of frequency in a sample. When a sample absorbs energy from the exciting source, the intensity of the absorption varies as a function of frequency. This variation is displayed as the absorption spectrum. The absorption spectrum is primarily determined by the atomic and molecular composition of the material [76, 77]. Absorbed frequencies match the quantum mechanical change of the material. Absorption is typically composed of multiple delta functions, but the shape of observed lines commonly have a Gaussian or Lorentzian shape that is determined by the instrument used for the observation, the material of the sample, its physical environment, the interactions between molecules in the sample, the crystal structure, and on ambient conditions like temperature, pressure, electrical fields, and magnetic fields.

Absorption lines may be classified by the nature of the quantum mechanical change induced in the molecules or atoms of a material.

- Rotational lines occur when the rotational state of a molecule is changed and are typically found in the microwave spectral region.
- Vibrational lines corresponding to changes in the vibrational state of the molecule are typically found in the infrared region.
- Electronic lines resulting from changes in the electronic state of an atom or molecule are typically found in the visible and ultraviolet region.
- X-ray absorptions are associated with the excitation of inner shell electrons in atoms.

Also, combinations of these changes such as rotation-vibration transitions, are present resulting in new absorption lines at the combined energy of the two changes.

A common mechanism for absorption measurements is to optically excite a sample and detect the radiation passing through it. The transmitted energy can be used to calculate the absorption. More elaborately, a reference spectrum of the source is first measured with a detector, and then, the sample is re-measured. The combination of the two measured spectra provides the material's absorption spectrum. The sample spectrum alone is not sufficient to determine the absorption spectrum because it will be affected by the experimental conditions. Thus, optical absorption spectroscopy acts as an important technique to study the absorption mechanism in materials and gives more insight into properties of low dimensional structures.

Device Fabrication

NCs have been the primary focus in the chapter to keep the explanations focused. Other types of NSs such as quantum dots, nanorods, and dye-sensitized NPs functionalized with quantum dots are also being investigated for similar applications. Freestanding NCs like the ones discussed in this chapter need to be embedded in a suitable matrix to fabricate devices. Amorphous Si, Silicon dioxide (SiO2), Zinc oxide (ZnO), Aluminum oxide (Al2O3) and organic polymers serve as matrix materials. Organic polymers provide the advantage of being able to synthesize flexible devices.

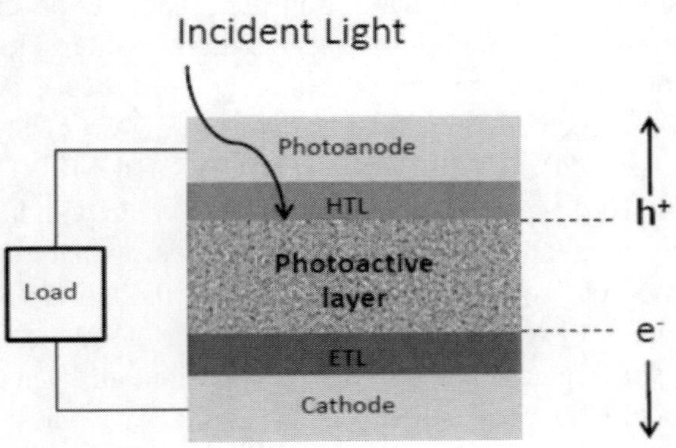

Figure 15. Nanocrystal-based PV Device.

The general structure of a typical NS-based PV device is as shown in figure 15 to sandwich the NS layer between two contacts. Additional layers are usually added between the electrodes and the active layer on each side, an electron transport layer that favors the passage of electrons and blocks holes on the cathode side and a hole-transport layer to favor the vice-versa in the anode side.

Light passes through the transparent anode and gets absorbed in the photoactive NS layer where carrier separation happens, giving rise to free electrons and holes. The electrons flow towards the cathode and the holes towards the anode. The active layer must facilitate exciton dissociation and subsequent carrier transport and the interfaces need optimal energies for the selective extraction of the carriers.

When photons with energy greater than the band gap enter the photoactive layer, they create an exciton, which is an electro-hole pair. Provided recombination of the electron and hole does not occur before the exciton reaches a encountering a suitable junction, the exciton dissociates into a free electron and a free hole. Following this, the carriers migrate to the respective electrodes under the influence of an internal electric field generated by a difference between the Fermi levels of the materials of the photoactive layer.

In case of the hybrid active layer consisting of a polymer and a semiconductor nanomaterial, researchers have used several approaches to synthesize the composite. One of the common approaches is to use the spin coating technique where the nanomaterials are dispersed in with the polymer and the stable dispersion is spun into a thin film using a spin caster [78, 79]. Another, more recent endeavor is to synthesize polymer nanofibers embedded with the nanomaterials using a variety of techniques such as chemical synthesis [80, 81, 82], nanolithography [83, 84], the physical drawing fabrication process [85, 86, 87, 88, 89, 90, 91], the high electric field electrospinning technique [92, 93, 94, 95, 96].

Introducing charge-transporting layers adjacent to cathode and anode such that each layer is favorable to the transport of one type of charge but blocks the other type, meaning an electron-transporter/hole-blocker on the side of the cathode and a hole-transporter/electron-blocker on the anode side could lead to solar cells with high power-conversion efficiencies. High efficiencies have been realized by substituting liquid electrolytes with solid hole-transporting materials [97]. Combined functionalities of light absorption and electron conduction have been realized by Lee et al. [98] using a mixed-halide perovskite on a mesoporous Al2O3 photoanode. Other researchers [99, 100, 101] have also demonstrated improved efficiencies with the use of charge transport layers. Band gap, ease of deposition, and chemical compatibility are important considerations while selecting carrier transport materials. Alq3 is a common electron transport material used for PV devices. Other available ETL materials developed include BCP, TBPI, TAZ, and BAlq [102]. Among a variety of HTL materials such as DPB, TTA, DEH, TPD, PTS, BCMF, TNF, and DPQ [103], PEDOT:PSS [104, 105, 106] is a common material used to help with hole transport.

For the contact layers metals such as Aluminum, Silver, palladium gold have been used for a long time. Transparent conducting oxides such as Indium tin oxide (ITO) have been popular [107, 108, 109, 110] as a contact material due to good electrical conductivity and optical transparency. The contact material may be deposited all over the cell rather than the opaque grid like structure in conventional cells. This allows for the complete solar cell surface to be exposed to the incident light through the transparent contact.

Lately, silver NWs are being actively researched [111, 112, 113, 114] as a contact material to obtain better flexibility and durability and improved transparency over ITO. Also,

ITO is not very compatible with polymers as it requires a high deposition temperature (600 °C). In addition, the expensive layer deposition process in vacuum and limited availability of indium results in high costs. More recently, conductive CNT coatings have become a potential substitute for ITO as various methods including spraying, spin coating, casting, layer-by-layer, and Langmuir–Blodgett process [115, 116, 117, 118] have been used for deposition. Graphene is another material under active investigation for PV cell contacts [119, 120] due to extreme conductivity, improved film-forming ability, complete transparency, cost-effectiveness, and nontoxic nature and highly promising results have been achieved for organic solar cells with grapheme contacts.

Device Characterization

Nanocrystal based solar cells are characterized using the same techniques as used for a conventional solar cell. Essentially the current-voltage characteristics are measured. Just as with conventional solar cell characterization, the V_{oc}, I_{sc}, efficiency, and fill factor provide sufficient information regarding the electrical behavior of the PV device under typical conditions. If the device is flexible, these measurements may be obtained with varying amounts of bending of the device to determine the effects of the strain on the performance of the device.

Conclusion

The above sections briefly describe the basics of NSs, synthesis processes and characterization techniques for NSs, and device fabrication and characterization. More details on these topics and information regarding specific NSs may be found in the books and publications listed in the references below as well as numerous other publications in this highly active area of research.

Acknowledgments

This work was supported in part by an appointment to the Postgraduate Research Participation Program at the U.S. Army Research Laboratory administered by the Oak Ridge Institute for Science and Education through an interagency agreement between the U.S. Department of Energy and USARL.

References

[1] S. Lo, Z. Liu, J. Li, H. L. Chan and F. Yan, "Hybrid solar cells based on poly(3-hexylthiophene) and electrospun TiO2 nanofibers modified with CdS nanoparticles," *Progress in Natural Science: Materials International,* vol. 23, no. 5, pp. 514-518, 2013.

[2] M. Beard, J. Luther, O. Semonin and A. Nozik, "Third generation Photovoltaics based on Multiple Exciton Generation in Quantum Confined Semiconductors," *Accounts of Chemical Research,* vol. 46, no. 6, pp. 1252-1260, 2012.

[3] S. G. Cloutier, P. A. Kossyrev and J. M. Xu, "Optical gain and stimulated emission in periodic nanopatterned crystalline silicon," *Nature Materials,* vol. 4, no. 12, pp. 887-891, 2005.
[4] E. Rotem, J. Shainline and J. Xu, "Electroluminescence of nanopatterned silicon with carbon implantation and solid phase epitaxial regrowth," *Optics Express,* vol. 15, no. 21, pp. 14099-14106, 2007.
[5] D. Recht, F. Capasso and M. J. Aziz, "On the temperature dependence of point-defect-mediated luminescence in silicon," *Applied Physics Letters,* vol. 94, no. 25, p. 251113, 2009.
[6] H. Rong, A. Liu, R. Jones, O. Cohen, D. Hak, R. Nicolaescu, A. Fang and M. Paniccia, "An all-silicon Raman laser," *Nature,* vol. 433, no. 7023, pp. 292-294, 2005.
[7] L. Pavesi, L. Negro, C. Mazzoleni, G. Franzò and F. Priolo, "Optical gain in silicon nanocrystals," *Nature,* vol. 408, no. 6811, pp. 440-444, 2000.
[8] W. L. Wilson, P. F. Szajowski and L. Brus, "Measurements of alloy composition and strain in thin GexSi1−x layers," *Science,* vol. 262, pp. 1242-1244, 1993.
[9] H. Enner, G. Pomrenke, A. Axmann, K. Eisele, W. Haydl and J. Schneider, "1.54-μm electroluminescence of erbium-doped silicon grown by molecular beam epitaxy," *Applied Physics Letters,* vol. 46, no. 4, pp. 381-383, 1985.
[10] A. W. Fang, H. Park, O. Cohen, R. Jones, M. J. Paniccia and J. E. Bowers, "Electrically pumped hybrid AlGaInAs-silicon evanescent laser," *Optics Express,* vol. 14, no. 20, pp. 9203-9210, 2006.
[11] A. E. Vasdekis, S. A. Moore, A. Ruseckas, T. Krauss, I. D. W. Samuel and G. Turnbull, "Silicon based organic semiconductor laser," *Applied Physics Letters,* vol. 91, p. 051124, 2007.
[12] L. Nataraj, N. Sustersic, M. Coppinger, L. F. Gerlein, J. Kolodzey and S. G. Cloutier, "Structural and optoelectronic properties of germanium-rich islands grown on silicon using molecular beam epitaxy," *Applied Physics Letters,* vol. 96, p. 121911, 2010.
[13] L. Nataraj, F. Xu and S. G. Cloutier, "Direct-bandgap luminescence at room temperature from highly-strained Germanium nanocrystals," *Optics Express,* vol. 18, no. 7, p. 7085, 2010.
[14] N. Sustersic, L. Nataraj, C. Weiland, M. Coppinger, M. V. Shaleev, A. V. Novikov, R. Opila, S. G. S. G. Cloutier and J. J. Kolodzey, "Effects of boron and phosphorus doping on the photoluminescence of self-assembled germanium quantum dots," *Applied Physics Letters,* vol. 94, p. 183103, 2009.
[15] L. Nataraj, A. Jackson, L. Giri, T. Atwater, C. Hubbard and M. Bundy, "Synthesis and Characterization of group-IV Semiconductor Nanocrystals," *IEEE Nanoscience and Nanotechnology Letters,* p. In Print, 2014.
[16] C. Kittel, Introduction to Solid State Physics, Wiley, 2004.
[17] A. Luque and A. Martí, "Increasing the Efficiency of Ideal Solar Cells by Photon Induced Transitions at Intermediate Levels," *Physical Review Letters,* vol. 78, p. 5014, 1997.
[18] A. Martí, L. Cuadra, N. López and A. Luque, "Intermediate band solar cells: Comparison with shockley-read-hall recombination," *Semiconductors,* vol. 38, no. 8, pp. 946-949, 2004.
[19] K.-Y. Ban, S. N. Dahal, C. B. Honsberg, L. Nataraj, S. P. Bremner and S. G. Cloutier, "Room temperature capacitance-voltage profile and photoluminescence for delta doped

InGaAs single quantum well," *Journal of Vacuum Science & Technology B,* vol. 28, no. 3, p. C316, 2010.

[20] E. F. Schubert, J. M. Kuo and R. F. Kopf, "Theory and Experiment of capacitance-Voltage Profiling on Semiconductors with Quantum Confinement," *Journal of Electronic materials,* vol. 19, no. 6, p. 521, 1990.

[21] L. Tsybeskov and D. J. Lockwood, "Silicon-Germanium anostructures for Light Emitters and On-Chip Optical Interconnects," *Proceedings of the IEEE,* vol. 97, no. 7, p. 1284, 2009.

[22] R. A. Soref, "Silicon-based Optoelectronics," in *Proceedings of the IEEE*, 1993.

[23] K. Wang, D. Cha, J. Liu and C. Chen, "Ge/Si Self-Assembled Quantum Dots and Their Optoelectronic Device Applications," *Proceedigns of the IEEE,* vol. 95, no. 9, p. 1866, 2007.

[24] M. L. Lee, E. A. Fitzgerald, M. Bulsara, M. Currie and A. Lochtefeld, "Strained Si, SiGe, and Ge channels for high-mobility metal-oxide-semiconductor field-effect transistors," *Journal of Applied Physics,* vol. 97, p. 3344, 2005.

[25] J. J. Welser, J. Hoyt and J. Gibbons, "Electron mobility enhancement in strained-Si N-type metal-oxide-semiconductor field-effect transistors," *IEEE Electron Device Letters,* vol. 15, no. 3, p. 100, 1994.

[26] D. Friedman, M. Meghelli, B. Parker, J. Yang, H. Ainspan, A. Rylyakov, Y. Kwark, M. Ritter, L. Shan, S. Zier, M. Sorna and M. Soyuer, "SiGe BiCMOS integrated circuits for high-speed serial communication links," *IBM Journal of Research and Development,* vol. 47, p. 259, 2003.

[27] J. Liu, X. Sun, D. Pan, X. Wang, L. Kimerling, K. T. L and J. Michel, "Tensile-strained, n-type Ge as a gain medium for monolithic laser integration on Si," *Optics Express,* vol. 15, no. 15, p. 11272, 2007.

[28] S. M. Hubbard, R. Raffaellea, R. Robinson, C. Bailey, D. Wilt, D. Wolford, W. Maurer and S. Bailey, "Growth and Characterization of InAs Quantum Dot Enhanced Photovoltaic Devices," in *Proceedings of MRS*, 2007.

[29] D. Kovalev, H. Heckler, G. Polisski and F. koch, "Optical Properties of Si Nanocrystals," *physia status solidi (b),* vol. 215, no. 2, p. 871, 1999.

[30] P. K. Giri, "Strain analysis on freestanding germanium nanocrystals," *Journal of Physics D: Applied Physics,* vol. 42, no. 24, p. 245402, 2009.

[31] A. Kanjilal, L. Rebohle, N. Baddela, S. Zhou, M. Voelskow, W. Skorupa and M. Helm, "Probing the impact of microstructure on the electroluminescence properties of Ge-nanocrystal enriched Er-doped SiO2 layers," *Physical Review B,* vol. 79, p. 161302, 2009.

[32] A. V. Kolobov, S. Q. Wei, W. S. Yan, H. Oyanagi, Y. Maeda and K. Tanaka, "Formation of Ge nanocrystals embedded in a SiO2 matrix: Transmission electron microscopy, x-ray absorption, and optical studies," *Physical Review B,* vol. 67, p. 195314, 2003.

[33] W. K. Choi, V. Ng, N. S. P, H. H. Thio, Z. X. Shen and W. S. Li, "Raman characterization of germanium nanocrystals in amorphous silicon oxide films synthesized by rapid thermal annealing," *Journal of Applied Physics,* vol. 86, p. 1398, 1999.

[34] W. Skorupa, L. Rebohle and T. Gebel, "Group-IV nanocluster formation by ion-beam synthesis," *Applied Physics A,* vol. 76, no. 7, p. 1049, 2003.

[35] H. -C. Weissker, J. Furthmüller and F. Bechstedt, "Optical properties of Ge and Si nanocrystallites from ab initio calculations. I. Embedded nanocrystallites," *Physical Review B,* vol. 65, p. 155327, 2002.

[36] S. Takeoka, M. Fujii, S. Hayashi and K. Yamamoto, "Size-dependent near-infrared photoluminescence from Ge nanocrystals embedded in SiO2 matrices," *Physical Review B,* vol. 58, p. 7921, 1998.

[37] I. D. Sharp, Q. Xu, C. Y. Liao, D. O. Yi, J. W. Beeman, Z. liliental-Weber, K. M. Yu, D. N. Zakharov, J. W. Ager III, D. C. Chrzan and E. E. Haller, "escholarship, University of California," Lawrence Berkley National Laboratory, 28 1 2005. [Online]. Available: http://www.escholarship.org/uc/item/65v92974,.

[38] Y. Maeda, "Visible photoluminescence from nanocrystallite Ge embedded in a glassy SiO2 matrix: Evidence in support of the quantum-confinement mechanism," *Physical Review Letters,* vol. 51, p. 1658, 1995.

[39] L. Dong, J. Schnitker, R. W. Smith and D. J. Srolovitz, "Stress relaxation and misfit dislocation nucleation in the growth of misfitting films: A molecular dynamics simulation study," *Journal of Applied Physics,* vol. 83, p. 217, 1998.

[40] M. Suezawa, Y. Sasaki and K. Sumino, "Dependence of Photoluminescence on Temperature in Dislocated Silicon Crystals," *physica status solidi (a),* vol. 79, no. 1, p. 173, 1983.

[41] R. Sauer, C. Kisielowski-Kemmerich and H. Alexander, "Dissociation-width-dependent radiative recombination of electrons and holes at widely split dislocations in silicon," *Physical Review Letters,* vol. 57, p. 1472, 1986.

[42] M. Milosavljević, M. A. Lourenço, G. Shao, R. M. Gwilliam and K. P. Homewood, "Optimising dislocation-engineered silicon light-emitting diodes," *Applied Physics B,* vol. 83, p. 289, 2006.

[43] N. Tit, N. Amrane and A. H. Reshak, "Comparison of Bowing Behaviors Between III–V and II–VI Common-Cation Semiconductor Ternary Alloys," *Journal of Electronic Materials,* vol. 39, no. 2, p. 178, 2010.

[44] Y.-T. Lin, T.-C. Ma, H.-H. Lin, J.-D. Wu and Y.-S. Huang, "Effect of thermal annealing on the blueshift of energy gap and nitrogen rearrangement in GaAsSbN," *Applied Physics Letters,* vol. 96, p. 011903, 2010.

[45] A. N. Pikhtin and H. H. Hegazy, "Fundamental absorption edge of semiconductor alloys with the direct-gap energy-band structure," *Semiconductors,* vol. 43, no. 10, p. 1259, 2009.

[46] X. Lu, D. A. Beaton, R. B. Lewis, T. Tiedje and Y. Zhang, "Composition dependence of photoluminescence of GaAs1−xBix alloys," *Applied Physics Letters,* vol. 95, p. 041903, 2009.

[47] J. Wu, "When group-III nitrides go infrared: New properties and perspectives," *Journal of Applied Physics,* vol. 106, p. 011101, 2009.

[48] M. Zacharias, J. Bläsing, J. Christen, P. Veit, B. Dietrich and D. Bimberg, "Formation of Ge nanocrystals with sharp size distribution: structural and optical characterization," *Superlattices and Microstructures,* vol. 18, no. 2, p. 139, 1995.

[49] Y. Maeda, T. N, Y. Yazawa, Y. Kanemitsu and Y. Masumoto, "Visible photoluminescence of Ge microcrystals embedded in SiO2 glassy matrices," *Applied Physics Letters,* vol. 59, p. 3168, 1991.

[50] Y. Kanemitsu, K. Masuda, M. Yamamoto, K. Kajiyama and T. Kushida, "Near-infrared photoluminescence from Ge nanocrystals in SiO2 matrices," *Journal of Luminescence,* Vols. 87-89, p. 457, 2000.

[51] Y. Kanemitsu, H. Uto, Y. Masumoto and Y. Maeda, "On the origin of visible photoluminescence in nanometer-size Ge crystallites," *Applied Physics Letters,* vol. 61, p. 2187, 1992.

[52] G. Kartopu, A. V. Sapelkin, V. A. Karavanskii, U. Serincan and R. Turan, "Structural and optical properties of porous nanocrystalline Ge," *Journal of Applied Physics,* vol. 103, no. 11, p. 113518, 2008.

[53] H. P. Wu, M. Y. Ge, C. W. Yao, Y. W. Wang, Y. W. Zeng, L. N. Wang, G. Q. Zhang and J. Z. Jiang, "Blue emission of Ge nanocrystals prepared by thermal decomposition," *Nanotechnology,* vol. 17, p. 5339, 2006.

[54] Q. Xiao and H. Tu, "Ge-Si system nanoclusters in Si matrix formed by solid-phase epitaxy," *Applied Physics Letters,* vol. 86, no. 20, p. 201914, 2005.

[55] H. Yang, X. Wang, H. Shi, S. Xie, F. Wang, X. Gu and X. Yao, "Photoluminescence of Ge nanoparticles embedded in SiO2 glasses fabricated by a sol–gel method," *Applied Physics Letters,* vol. 81, p. 5144, 2002.

[56] E. Palange, G. Capellini, L. Di Gaspare and E. F, "Atomic force microscopy and photoluminescence study of Ge layers and self-organized Ge quantum dots on Si(100)," *Applied Physics Letters,* vol. 68, p. 2982, 1996.

[57] S. W. Lee, P. S. Chen, S. L. Cheng, M. H. Lee, H. T. Chang, C. -H. Lee and C. W. Liu, "Modified growth of Ge quantum dots using C2H4 mediation by ultra-high vacuum chemical vapor deposition," *Applied Surface Science,* vol. 254, p. 6261, 2008.

[58] D. C. Paine, C. Caragianis, T. Kim, Y. Shigesato and T. Ishahara, "Visible photoluminescence from nanocrystalline Ge formed by H2 reduction of Si0.6Ge0.4O2," *Applied Physics Letters,* vol. 62, p. 2842, 1993.

[59] S. Miyazaki, K. Sakamoto, K. Shiba and M. Hirose, "Photoluminescence from anodized and thermally oxidized porous germanium," *Thin Solid Films,* vol. 255, p. 99, 1995.

[60] D. -H. Choi, Y. D. Han, B. -K. Lee, C. S. -J, Y. H. C. C, L. D. -S and J. -B. Yoon, "Use of a Columnar Metal Thin Film as a Nanosieve with Sub-10 nm Pores," *Advanced Materials,* vol. 24, p. 4408, 2012.

[61] D. K. Schroder, Semiconductor Material and Device Characterization, Wiley, 2006.

[62] D. R. Wolters and J. J. Van der Schoot, " 40, 115 (1985).," *Philips Research Reports,* vol. 40, p. 115, 1985.

[63] J. S. Siegert, L. Fu and C. Jagadish, "Recombination properties of Si-doped InGaAs/GaAs quantum dots," *Nanotechnology,* vol. 17, no. 21, p. 5373, 2006.

[64] T. Schmidt, K. Lischka and W. Zulehner, "T. Schmidt et al. 45, 16 (1992).," *Physical Review B,* vol. 45, p. 8989, 1992.

[65] T. Taguchi, J. Shirafuji and Y. Inuishi, "Excitonic emission in cadmium telluride," *Physica Status Solidi B 68, 727 (1975),* vol. 68, no. 2, p. 727, 1975.

[66] D. E. Cooper, J. Bajaj and P. R. Newman, "Photoluminescence spectroscopy of excitons for evaluation of high-quality CdTe crystals," *Journal of Crystal Growth,* vol. 86, p. 544, 1988.

[67] Z. C. Feng and A. Mascarenhas, "Low temperature photoluminescence spectra of (001) CdTe films grown by molecular beam epitaxy at different substrate temperatures," *Journal of Luminescence,* vol. 35, no. 6, p. 329, 1986.

[68] Q. Kim and D. W. Langer, "Effects of Excitation Intensity on Photoluminescence of Pure CdTe," *Physica Status Solidi B* 122, 263 (1984)., vol. 122, no. 1, p. 263, 1984.

[69] H. Shibata, M. Sakai, A. Yamada, K. Matsubara, K. Sakurai, H. Tampo, S. Ishizuka, K.-K. Kim and S. Niki, "Excitation-Power Dependence of Free Exciton Photoluminescence of Semiconductors," *Japanese Journal of Applied Physics,* vol. 44, no. 8, p. 6115, 2005.

[70] I. Schnitzer and E. Yablonovitch, "30% External Quantum Efficiency from Surface-textured, thin film Light Emitting Diodes," *Applied Physics Letters* 63, 2174 (1993), vol. 63, no. 16, p. 2174, 1993.

[71] S. G. Cloutier, P. A. Kossyrev and J. M. Xu, "Phonon-localization in periodic uniaxially-nanostructured silicon," *Applied Physics Letters,* vol. 87, p. 222104, 2005.

[72] D. C. D. C. Harris and M. D. Bertolucci, Symmetry and Spectroscopy, Dover, 1978.

[73] O. V. Vakulenko, S. V. Kondratenko, A. S. Nikolenko, S. L. Golovinskiy, Y. N. Kozyrev, M. Y. Rubezhanska and A. I. Vodyanitsky, "Photoconductivity spectra of Ge/Si heterostructures with Ge QDs," *Nanotechnology,* vol. 18, p. 185401, 2007.

[74] P. W. Fry, I. E. Itskevich, S. R. Parnell, J. J. Finley, L. R. Wilson, K. L. Schumacher, D. J. Mowbray, M. S. Skolnick, M. Al-Khafaji, A. G. Cullis, M. Hopkinson, J. C. Clark and G. Hill, "Photocurrent spectroscopy of InAs/GaAs self-assembled quantum dots," *Physical Review B* 62, 24 (2000), vol. 62, p. 24, 2000.

[75] T. P. Pearsall, H. Polatoglou, H. Presting and E. Kasper, *Physical Review B* 54, 1545 (1996), vol. 54, p. 1545, 1996.

[76] D. C. Harris and M. D. Bertolucci, Symmetry and Spectroscopy, 1978.

[77] J. M. Hollas, Modern Spectroscopy, 2004.

[78] R. Zhou, R. Stalder, D. Xie, W. Cao, Y. Zheng, Y. Yang, M. Plaisant, P. H. Holloway, K. S. Schanze, J. R. Reynolds and J. Xue, "Enhancing the Efficiency of Solution-Processed Polymer:Colloidal Nanocrystal Hybrid Photovoltaic Cells Using Ethanedithiol Treatment," *Nano,* vol. 7, no. 6, p. 4846, 2013.

[79] W. C. Liu, A. D. Li, C. L. Mak and K. H. Wong, "Fabrication and Electro-optic Properties of Ferroelectric Nanocrystal/Polymer Composite Films," *Journal of Physical Chemistry,* vol. 112, no. 36, p. 14202, 2008.

[80] L. W and M. C. R, "Template-synthesized polyacetylene fibrils show enhanced supermolecular order," *Journal of Americal Chemical Society,* vol. 112, no. 26, p. 9666, 1990.

[81] Z. Cai, J. Lei, W. Liang, V. Menon and C. Martin, "Molecular and supermolecular origins of enhanced electronic conductivity in template-synthesized polyheterocyclic fibrils. 1. Supermolecular effects.," *Chemistry of Materials,* vol. 3, p. 960, 1991.

[82] T. Cui, F. Cui, J. Zhang, J. Wang and J. Huang, "From monomeric nanofibers to PbS nanoparticles/polymer composite nanofibers through the combined use of γ-irradiation and gas/solid reaction," *Journal of American Chemical Society* 2006; 128: 6298–6299, vol. 128, no. 19, p. 6298, 2006.

[83] A. Noy, M. A.E, K. J.E, W. B.L and W. B.W, "Fabrication of luminescent nanostructures and polymer nanowires using dip-pen nanolithography," *Nano Letters,* vol. 2, p. 109, 2002.

[84] C. D. Marco, E. Mele, A. Camposeo, R. Stabile and C. R, "Organic light-emitting nanofibers by solvent-resistant nanofluidics.," *Adv Mater* 2008; 20: 4158–4162., vol. 20, p. 4158, 2008.

[85] H. S.A, Cambron S.D, N. E.W, S. Berry and A. Isham, "Direct drawing of suspended filamentary micro- and nanostructures from liquid polymers.," *Nano Letters,* vol. 4, p. 1931, 2004.

[86] F. Gu, L. Zhang, X. Yin and L. Tong, "Polymer single-nanowire optical sensors," *Nano Letters,* vol. 8, p. 2757, 2008.

[87] X. Xing, H. Zhu, Y. Wang and B. Li, "Ultracompact photonic coupling splitters twisted by PTT nanowires," *Nano Letters,* vol. 8, p. 2839, 2008.

[88] C. Meng, Y. Xiao, P. Wang, L. Zhang and Y. Liu, "Quantum-dot-doped polymer nanofibers for optical sensing," *Adv Mater* 2011; 23: 3770–3774, vol. 23, p. 3770, 2011.

[89] P. Wang, L. Zhang, Y. Xia, L. Tong and X. Xu, "Polymer nanofibers embedded with aligned gold nanorods: a new platform for plasmonic studies and optical sensing.," *Nano Letters 2012; 12: 3145–3150,* vol. 12, p. 3145, 2012.

[90] F. Gu, H. Yu, W. Fang and L. Tong, "Nanoimprinted polymer micro/nanofiber Bragg gratings for high-sensitivity strain sensing," *IEEE Photon Technology Letters,* vol. 25, p. 22, 2013.

[91] F. Benedetto, E. Mele, A. Camposeo, A. Athanassiou and R. Cingolani, "Photoswitchable organic nanofibers," *Advanced Materials,* vol. 20, p. 314, 2008.

[92] Y. Zhang, Y. Liu, X. Li, Q. Wang and E. Xie, "Room temperature enhanced red emission from novel Eu(3+) doped ZnO nanocrystals uniformly dispersed in nanofibers.," *Nanotechnology,* vol. 22, no. 41, p. 415702, 2011.

[93] D. Li and Y. Xia, "Electrospinning of nanofibers: reinventing the wheel?," *Advanced Materials,* vol. 16, p. 1151, 2004.

[94] A. Camposeo, F. D. Benedetto, R. Stabile, A. A. R. Never and R. Cingolani, "Laser emission from electrospun polymer nanofibers," *Small,* vol. 5, p. 562, 2009.

[95] H. Liu, J. Edel, L. Bellan and H. Craighead, "Electrospun polymer nanofibers as subwavelength optical waveguides incorporating quantum dots.," *Small,* vol. 2, p. 495, 2006.

[96] Y. Dzenis, "Spinning continuous fibers for nanotechnology," *Science,* vol. 304, p. 1917, 2004.

[97] O. Malinkiewicz, A. Yella, Y. H. Lee, G. M. Espallargas, M. Graetzel, M. K. Nazeeruddin and H. J. Bolink, "Perovskite solar cells employing organic charge-transport layers," *Nature Photonics,* vol. 8, p. 128, 2014.

[98] M. M. Lee, J. Teuscher, T. Miyasaka, T. N. Murakami and H. J. Snaith, *Science 338,* vol. 338, p. 643, 2012.

[99] E. Danielson, H. Subbaraman and A. Dodabalapur, "Inkjet printing of carrier transport layers for inverted organic solar cells," in *SPIE,* Volume 8819, 2013.

[100] H. Hoppe and N. S. Sariciftci, "Organic solar cells: An overview," *Journal of Materials Research,* vol. 19, no. 7, p. 1924, 2004.

[101] B. Tremolet De Villers, C. J. Tassone, S. H. Tolbert and B. J. Schwartz, "Improving the Reproducibility of P3HT:PCBM Solar Cells by Controlling the PCBM/Cathode Interface," *The Journal of Physical Chemistry C* 113 (44): 18978, vol. 113, no. 44, p. 18978, 2009.

[102] L. Cosimbescu, L. Wang, M. L. Helm, E. Polikarpov, J. S. Swensen and A. B. Padmaperuma, "Electron Transport Materials: Synthesis, Properties and Device Performance," *International Journal of Organic Chemistry,* vol. 2, p. 101, 2012.

[103] S. Edited by Kasap and P. Capper, Springer Handbook of Electronic and Photonic Materials, Springer Science, 2006.
[104] D. Wakizaka, T. Fushimi, H. Ohkita and S. Ito, "Hole transport in conducting ultrathin films of PEDOT/PSS prepared by layer-by-layer deposition technique," *Polymer,* vol. 45, no. 25, p. 8561, 2004.
[105] D. Bailo, A. Generosi, V. R. Albertini, R. Caminiti, R. De Bettignies and B. Paci, "Time-resolved morphological study of 'PEDOT:PSS' hole transporting layer for polymer solar cells," *Synthetic Metals,* vol. 162, no. 9-10, p. 808, 2012.
[106] F. J. Lim, K. Ananthanarayanan, J. Luther and G. W. Ho, "Influence of a novel fluorosurfactant modified PEDOT:PSS hole transport layer on the performance of inverted organic solar cells," *Journal of Materials Chemistry,* vol. 22, p. 25057, 2012.
[107] G. G. Pethuraja, R. E. Welser, A. K. Sood, C. Lee, N. J. Alexander, H. Efstathiadis, P. Haldar and J. L. Harvey, "Current-Voltage Characteristics of ITO/p-Si and ITO/n-Si Contact Interfaces," *Advances in Materials Physics and Chemistry,* vol. 2, p. 59, 2012.
[108] Q. Wang, M. R. Page, E. Iwaniczko, Y. Xu, L. Roybal, R. Bauer, B. To, H. C. Yuan, A. Duda and Y. F. Yan, "Crystal Silicon Heterojunction Solar Cells by Hot-Wire CVD," in *IEEE Photovoltaic Specialists Conference*, San Diego, 2008.
[109] Z. Eygi, U. Das, S. Hegedus and R. Birkmire, "Improved FF in P-Si heterojunction solar cells due to optimized ITO/emitter contact," in *37th IEEE Photovoltaic Specialists Conference (PVSC)*, Seattle, 2011.
[110] Y. Cheng, Z. Gang, L. Dae-Young, L. Hua-Min, L. Young-Dae, Y. W. Jong, P. Young-Jun and K. Jong-Min, "Self-Assembled Wire Arrays and ITO Contacts for Silicon Nanowire Solar Cell Applications," *Chinese Physics Letters,* vol. 28, p. 035202, 2011.
[111] J. H. Yim, S.-Y. Joe, C. Pang, K. M. Lee, H. Jeong, J.-Y. Park, Y. H. Ahn, J. C. de Mello and S. Lee, "Fully Solution-Processed Semitransparent Organic Solar Cells with a Silver Nanowire Cathode and a Conducting Polymer Anode," *Nano,* 2014.
[112] M. Song, D. S. You, K. Lim, S. Park, S. Jung, C. S. Kim, D.-H. Kim, D.-G. Kim, J.-K. Kim, J. Park, Y.-C. Kang, J. Heo, S.-H. Jin, J. H. Park and J.-W. Kang, "Highly Effi cient and Bendable Organic Solar Cells with Solution-Processed Silver Nanowire Electrodes," *Advanced Functional Materials,* vol. 23, p. 4177, 2013.
[113] D.-S. Leem, A. Edwards, M. Faist, J. Nelson, D. C. Bradley and J. C. de Mello, "Efficient Organic Solar Cells with Solution-Processed Silver Nanowire Electrodes," *Advanced Materials* Iss23, 38, 4371, vol. 38, no. 23, p. 4371, 2011.
[114] M. A. Hines and G. D. Scholes, "Colloidal PbS Nanocrystals with Size-Tunable Near-Infrared Emission: Observation of Post-Synthesis Self-Narrowing of the Particle Size Distribution," *Advanced Materials,* vol. 15, no. 21, p. 1844, 2003.
[115] H. Ago, K. Petritsch, M. Shaffer, A. H. Windle and R. H. Friend, "Composites of carbon nanotubes and conjugated polymers for photovoltaic devices," *Advanced Materials,* vol. 11, no. 15, p. 1281, 1999.
[116] M. Contreras, T. Barnes, J. Van De Lagemaat, G. Rumbles, T. Coutts, C. Weeks, P. Glatkowski, I. Levitsky, J. Peltola and D. Britz, "Replacement of transparent conductive oxides by single-wall carbon nanotubes in Cu(In,Ga)Se2-based solar cells," *Journal of Physical Chemistry C,* vol. 111, no. 38, p. 14045, 2007.
[117] T. V. Sreekumar, T. Liu, S. Kumar, L. Ericson, R. Hauge and R. Smalley, "Single-wall carbon nanotube films," *Chemistry of Materials,* vol. 15, no. 1, p. 175, 2003.

[118] X. Li, L. Zhang, X. Wang, I. Shimoyama, X. Sun, W.-K. Seo and H. Dai, "Langmuir-Blodgett assembly of densely aligned single-walled carbon nanotubes from bulk materials," *Journal of the American Chemical Society* 129 (16): 4890–4891, vol. 129, no. 16, p. 4890, 2007.

[119] W. Xuan, Z. Linjie and M. Klaus, "Transparent, Conductive Graphene Electrodes for Dye-Sensitized Solar Cells," *Nano Letters,* vol. 8, no. 1, p. 323, 2008.

[120] H. Wenjing, X. Yuxi, L. Gewu, L. Chun and S. Gaoquan, "Transparent graphene/PEDOT–PSS composite films as counter electrodes," *Electrochemistry Communications,* vol. 10, p. 1555, 2008.

In: Photovoltaics
Editor: Monique A. Gill, pp. 37-51
ISBN: 978-1-63117-843-6

Chapter 2

EVALUATION OF POTENTIAL HEALTH AND ENVIRONMENTAL IMPACTS FROM END-OF-LIFE DISPOSAL OF PHOTOVOLTAICS

Parikhit Sinha*[1,*]*, V. Lyle Trumbull*[2]*,
***Swiatoslav W. Kaczmar*[3]* and Keith A. Johnson*[4]**
[1]First Solar, Tempe, AZ, US
[2]O'Brien & Gere, E. Norriton, PA, US
[3]SUNY Upstate Medical University, Syracuse, NY, US
[4]Tetra Tech BAS, Phoenix, AZ, US

Abstract

Global solar photovoltaic (PV) deployment has enjoyed exponential growth in the past decade and as the industry grows, large future end-of-life PV waste volumes will need to be managed. It is likely that recycling will be a dominant strategy as in the case of the European Union, where end-of-life collection and recycling will be mandatory beginning 2014 under the recast Waste from Electrical and Electronic Equipment Directive. However, PV deployment is globalizing beyond the European Union into emerging markets. Given the presence of various metals in PV modules, some stakeholders have raised concerns regarding potential environmental impacts if PV modules are disposed of in unlined landfills instead of sanitary landfills or instead of being recycled. Potential adverse health or environmental impacts associated with disposal of utility-scale (25 MWac) volumes of PV modules in an unlined landfill were evaluated by fate and transport modeling using U.S. EPA Delisting Risk Assessment Software (DRAS). Estimated surface and groundwater media concentrations of Cd and Pb, and associated cancer risks and non-cancer hazards were below human health and ecological screening limits. The hazard index under basic landfill conditions was an order of magnitude lower than under acidic conditions, which is notable as landfills have predominantly neutral to slightly basic conditions over their lifetime. Estimated environmental concentrations of Pb from crystalline silicon PV were comparable to concentrations of Cd

[*]E-mail address: parikhit.sinha@firstsolar.com.

from cadmium telluride PV, indicating that responsible disposal is important for all PV technologies. Potential impacts may be further limited by use of sanitary landfills or with high value recycling, with the latter providing the important benefit of resource recovery. The DRAS default assumption is Toxicity Characteristic Leachate Procedure (TCLP) data for waste with a fragment size of 1 cm. Experimental crushing of cadmium telluride PV modules in landfill conditions was conducted to evaluate the representativeness of the leachate data used in the DRAS modeling. On average approximately three-quarters of the crushed module fragments were larger than 1 cm, and 99% of module fragments were larger than 0.1 mm. The glass-laminate-glass bond of individual broken module fragments was maintained, and when analyzed without further size reduction, the landfill fragments are non-hazardous based on TCLP and Soluble Threshold Limit Concentration (STLC) testing.

Introduction

Global solar PV deployment has enjoyed exponential growth in the past decade. As the industry grows, PV module recycling is important to managing large future end-of-life waste volumes and conserving raw materials for use in new PV modules. In the European Union (EU), where global PV deployment has historically been concentrated, end-of-life collection and recycling will be mandatory as of 2014, resulting from the inclusion of PV in the recast Waste from Electrical and Electronic Equipment (WEEE) Directive [1]. However, PV deployment is globalizing beyond the EU into emerging markets where a combination of high solar irradiation and unmet and growing demand for electricity can create economically sustainable markets for PV. In these new markets, end-of-life collection and recycling may be achieved voluntarily through PV manufacturers or third party services, or may be mandated in future WEEE-type regulations or through project decommissioning requirements.

It is likely that recycling will be the dominant strategy used to manage future end-of-life waste volumes of PV modules. However, given the presence of various metals in silicon (Si) and cadmium telluride (CdTe) PV modules, some stakeholders have raised concerns regarding potential environmental impacts if PV modules are disposed of in unlined landfills instead of sanitary landfills or instead of being recycled. This concern has been evaluated in this study using a combination of experimental analysis and fate and transport modeling with the U.S. EPA Delisting Risk Assessment Software (DRAS) [2].

In the case of the metal Cd which occurs naturally in Zn, Pb and Cu ores and in coal and other fossil fuels, major industrial releases have been due to waste streams and leaching of landfills, and from a variety of operations that involve cadmium or zinc [3]. For example, there were well documented releases of Cd from Zn mining and smelting operations in the Toyama Prefecture of Japan in the early to mid-twentieth century. With respect to landfills, the disposal of NiCd batteries has been reported to have caused an increase of Cd concentrations in municipal solid waste landfills [4]. Cadmium content in NiCd batteries is in the form of Cd and $Cd(OH)_2$, materials which are less stable and more soluble than the CdTe semiconductor material considered in this study [5].

Potential impacts from landfill disposal of CdTe PV modules have previously been evaluated by Tvermoes et al. using DRAS [6]. Assumptions applied by Tvermoes et al. include landfilling of a large initial waste volume (5,600 kg Cd – corresponding to an 80 MWac project), and continuous disposal of CdTe modules over the active life of landfill (i.e., 1,225 kg Cd per year from years 2 through 20 corresponding to a total of approximately 350 MWac). Other conservative assumptions built into the DRAS model are that the landfill is unlined and is operating under acidic (fermentation) conditions for the life of the landfill. In reality, landfills have predominantly neutral to slightly basic (methanogenic) conditions over their lifetime [7]. Under these conditions, metal ions are rendered immobile by the formation of metal sulfides. Metal mobility in landfills is also limited by physical mechanisms such as sorption onto clays, silts, and ferrous complexes.

In this evaluation, the DRAS model was used to provide an estimate of potential health and environmental impacts associated with the disposal of end-of-life PV modules in unlined landfills, under conditions more typical and representative of existing and planned utility-scale PV projects. The evaluated scenarios are based on a one-time disposal of a waste volume corresponding to a median size utility-scale PV project, 25MWac, and consider Si PV modules that utilize Pb solder as well as CdTe PV modules. Note that although this study considers utility-scale PV, the waste volume considered is equivalent to that from numerous smaller scale PV systems (e.g., one thousand 25 kW systems). In addition, this study considers the influence of increases in pH that typically take place within a landfill over time with the progression of short-lived fermentation conditions to long-lived methanogenic conditions which limit the mobilization of metals from a typical PV module. Leachate concentrations, which are key inputs to the DRAS model, are also evaluated in this study through a landfill experiment. In sum, the objective of this study is to investigate the potential impacts to human health and the environment if local conditions result in PV landfill disposal instead of recycling with consideration of the fermentation and methanogenic landfill phases and unlined landfills.

Methods

The DRAS model evaluates risk and hazard for several groundwater exposure scenarios (ingestion of groundwater, dermal absorption while bathing with groundwater, and inhalation of groundwater volatiles while showering) and for three surface exposure pathways (ingestion of surface water, ingestion of surface soil, ingestion of fish, and inhalation of vapor and particles).

Two types of DRAS model runs were conducted to investigate the relationship between landfill chemistry and potential emissions from landfilled PV modules, a model representing an acidic (fermentation) landfill phase, and a model representing a neutral to basic (methanogenic) landfill phase. The only difference between these model runs was the type of leachate data used, with acidic conditions using the results of TCLP testing (sodium acetate/acetic acid solvent) [8] and neutral to basic conditions using the results of column testing (deionized water solvent) [9]. The basic pH conditions from use of deionized water solvent are due to the release of sodium from the soda lime glass surface of the PV module sample. TCLP and column test parameters are summarized in Table 1. The TCLP test results were commissioned by First Solar and the column testing was performed by Norwegian Geotechnical Institute.

All other DRAS parameters were the same for the two model runs. For the waste management unit type, the “landfill” option was selected.

Table 1. Waste characterization test parameters for acidic and basic landfill conditions

		Acidic Landfill Conditions	Basic Landfill Conditions
Leachate Test		TCLP [8]	Column [9]
Fragment Size (cm)		1.0	0.4
pH	CdTe PV	2.88 for alkaline waste; 4.93 for neutral to acidic waste	8.0 - 10.3
	Si PV		8.1-9.8
Solvent		sodium acetate/ acetic acid	deionized water
Test Method		end-over-end agitation	water percolates through the sample
Test Duration		18 hr	3 weeks
Leachate Cd Concentration (mg/L)	CdTe PV	<1	<0.002
	Si PV	<0.1	0.007
	Limit[a]	1	0.3
Leachate Pb Concentration (mg/L)	CdTe PV	0.005	0.0025
	Si PV	3-11 (median <5)	0.008
	Limit[a]	5	3

[a]Regulatory limit for non-hazardous waste in the U.S. (TCLP) and EU (column test).
TCLP = Toxicity Characteristic Leachate Procedure.

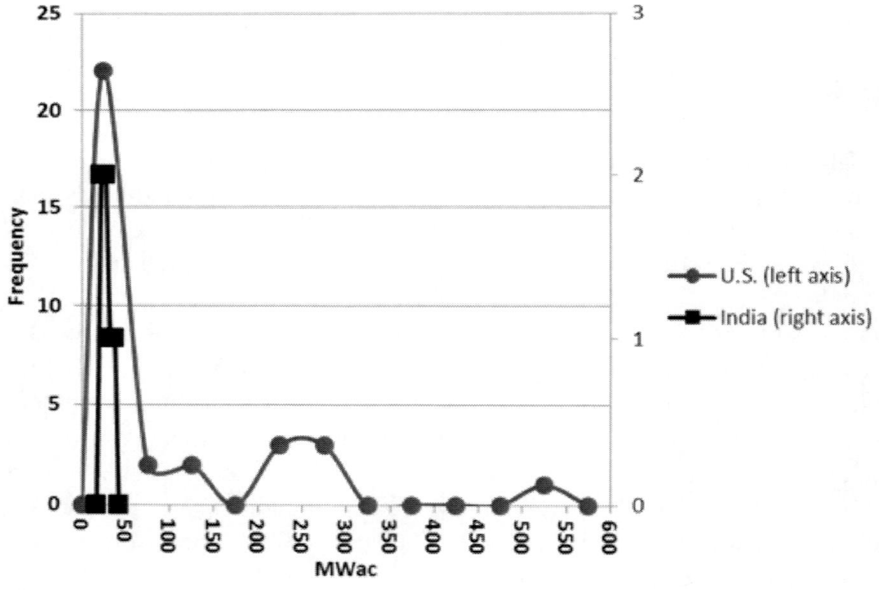

Figure 1. Utility-scale CdTe PV projects in developed and emerging markets.

Waste volume was set at 2,118 cubic yards, corresponding to a median size utility-scale PV project, 25 MWac [10] (Figure 1). For the waste management unit's active life, the "one year batch" option was selected as it is assumed that the entire 25 MWac project will decommissioned during one calendar year. In addition to the leachate concentrations for the acidic and basic scenarios provided in Table 1, the total constituent concentrations used were 383 mg Cd/kg and 15 mg Pb/kg for CdTe PV and 576 mg Pb/kg for Si PV [9].

Cd concentrations in soil and surface water are not explicitly provided in DRAS model output but they were derived from back-calculations using the DRAS-provided hazard quotients and media-specific hazard equations (equation 4-95 for surface water and equation 4-112 for soil in [2]). The modeled soil Cd concentration was compared to the lowest U.S. EPA Ecological Soil Screening Level for Cd [11]. Modeled surface water concentrations were compared to the U.S. EPA National Recommended Water Quality Criteria for Aquatic Life (Freshwater/Chronic) [12].

Exposure of ecological receptors to groundwater and ambient air were not evaluated in this assessment as it is standard ecological risk assessment practice to consider exposure via these pathways to be de minimis [2].

The methodology used to calculate hazards from the Pb concentrations that would result from landfilled Si and CdTe PV modules was different from the above-described methodology used for Cd. This is because toxicity factors (i.e., reference doses and cancer slope factors) are not available for Pb, as human health effects from Pb exposure may occur at blood lead levels so low as to be essentially without a threshold [13]. In this study, the potential for adverse effects to human health and the environment associated with Pb exposure were evaluated by comparing modeled media concentrations to medium-specific human health and ecological screening levels. This approach is consistent with DRAS guidance [2]. As with Cd, Pb concentrations for the relevant environmental media were not provided by DRAS and were derived as follows.

The concentration of Pb in groundwater was calculated using the DRAS-provided waste volume adjusted dilution attenuation factor (DAF) and by rearranging DRAS Equation 2-1 in [2] to solve for the waste constituent concentration in groundwater. Pb concentrations from the DRAS surface pathway (surface water, ambient air, and soil) were derived using a similar methodology which is based on the two DRAS input parameters that influence these media concentrations, waste volume and total constituent concentration in waste. The waste volume was held constant at 2118 yd^3 for all DRAS model runs. The total constituent concentrations in waste were 15 mg Pb/kg for CdTe PV and 576 mg Pb/kg for Si PV [9].

Once calculated, DRAS guidance recommends that the Pb concentrations are compared to the following human health-based levels: 0.015 mg/L for drinking water, 100 mg/kg for soil, and 0.2 $\mu g/m^3$ for ambient air [2]. DRAS sets the soil and air target levels at 25 percent of the actual risk-based levels based on the assumption that only 25 percent of the risk-based level is attributable to releases from the landfill modeled in DRAS. Ecological exposure to Pb was evaluated as described above for Cd, with comparison to Ecological Soil Screening Level for Pb [14], National Recommended Water Quality Criteria for surface water [12], and no ecological evaluation of groundwater or ambient air.

To experimentally evaluate the representativeness of the TCLP leachate data used with the DRAS model, five CdTe PV modules were crushed with a compactor at the top of a Municipal Solid Waste Landfill in the State of Arizona, USA. The modules were First Solar Series 3 Model 387 and were 0.6 m by 1.2 m in size, weighing approximately 12 kg. In the module design, the semiconductor layer was encapsulated between two sheets of protective glass, and a polymeric adhesive was used to laminate the two sheets of glass together.

The compactor used was a sheep's foot roller manufactured by Aljon, model 91K, with Caron cleats having a contact load of approximately 45,000 kg (Figure 2). The PV modules were placed on top of a 7 m × 12 m tarp consisting of a 1.5 mm thick linear low density polyethylene (LLDPE) liner. The compactor made six passes over the test installation. The ground surface underlying the test installation consisted of loose soil which was firmer than the typical mixed waste conditions at the active face of a landfill. The residue from the crushed PV modules was collected and separately bagged for testing.

The procedure used to crush the PV modules may be considered a worst case for module loading as the landfill's compactor is one of the largest and heaviest compactors available on the market. Landfills outside the United States typically do not have compaction equipment, and if landfill compaction equipment is available, the equipment is typically a track mounted dozer rather than a compactor. A dozer is typically used when the landfill can only afford one piece of equipment because the dozer can push the waste as well as compact the waste once it has been pushed into the active face. A track mounted dozer would be lighter than the landfill's compactor and would therefore cause less damage to the PV modules.

Figure 2. Aljon model 91K compactor used to crush CdTe PV modules in a Municipal Solid Waste Landfill in the State of Arizona, USA.

After the PV modules were crushed, the resultant fragments were sized using gradation assessment methodologies. The crushed PV modules consisted of fragments that were not attached to the module as well as larger pieces that were held together by the interior lamination. The front glass was observed to be crushed more than the back glass. It appeared that the material fractures followed the stress pattern imparted by the compactor feet (Figure 3).

The gradation assessment adopted the particle count procedure of the U.S. Bureau of Reclamation known as a "Pebble Count" on a six inch by six inch section of crushed modules. The crushed fragments attached to the modules were counted with an assignment of maximum dimension of 13 mm, as most of the fragments were smaller than 13 mm minimum dimension. Some fragments were elongated and the dimensions in longer axis were in some cases much larger than a few inches. The residual fragments were subjected to the American

Society for Testing and Materials (ASTM) method specified in ASTM D 422 without hydrometer analysis.

After gradation testing was completed, a representative sample of the PV modules was selected from bag #2 (PV Module 2) and bag #5 (PV Module 5). This representative sample consisted of (by weight) approximately two thirds intact module (crushed but in one piece) and one third loose fragments. In both cases, the glass-laminate-glass bond of individual broken module fragments was maintained (i.e., individual module fragments remained laminated). PV Modules 2 and 5 were chosen on the basis of their location in the test installation (in the center and perimeter, respectively).

The samples were sent for waste characterization testing based on the TCLP waste characterization test used federally in the U.S. and the STLC test used in California. Once the samples were received by the laboratory they were analyzed without any further reduction in size. TCLP test parameters for the landfill fragments were the same as the standard TCLP test methods in Table 1 except for the particle size. STLC testing requires a longer test duration (48 hr) compared with TCLP (18 hr) and uses end-over-end agitation in a sodium citrate solution (pH 5).

Results

Based on fate and transport modeling with the U.S. EPA DRAS model, the one-time disposal of CdTe PV modules from a 25 MWac project to an unlined landfill is not likely to represent significant cancer risks or non-cancer hazards (Table 2).

The aggregate cancer risk for both the acidic and basic scenarios (4.49×10^{-9}) was well below the screening limit of 1×10^{-6}. Likewise, the aggregate non-cancer hazard indices for the acidic (4.55×10^{-2}) and the basic (1.21×10^{-3}) landfill stages were well below the screening limit of 1, with hazard quotients for the basic landfill condition an order of magnitude lower than under acidic conditions. Potential adverse impacts to ecological receptors are unlikely as modeled surface water and soil Cd concentrations were 2-4 orders of magnitude below their respective ecological criteria.

Similar to the results for Cd, modeled Pb concentrations indicate that the one-time disposal of either CdTe PV modules or Si PV modules from a utility-scale project (2,118 cubic yards of modules) to an unlined landfill is not likely to represent significant risks or hazards to human or ecological receptors (Table 3). Modeled media concentrations of Pb from Si PV modules were one to three orders of magnitude larger than those modeled from CdTe PV modules but all media concentrations were below human health and ecological criteria.

A key input to the DRAS model was the TCLP leachate concentration. Experimental crushing of CdTe PV modules in landfill conditions was conducted to evaluate the representativeness of the TCLP data used in the DRAS modeling. After crushing, the PV modules largely held together with separate broken pieces only occurring where impacted by one of the landfill compactor's feet, which created a punch-out in the module of the approximate size of the foot. The PV module material punched-out of the modules was recovered to the extent possible and consisted of the finer fraction of the overall sample (Figure 3).

Table 2. Human health and ecological screening assessment of Cd from landfill disposal of utility-scale CdTe PV modules

	Pathway	*Exposure Scenario*	*Receptor*	*Cancer Risk*		*Hazard Quotient*	
				Acidic Landfill Conditions	**Basic Landfill Conditions**	**Acidic Landfill Conditions**	**Basic Landfill Conditions**
Human Health	Surface Pathway[a]	Inhalation of particles	Adult Resident	4.49×10^{-9}	4.49×10^{-9}	---	---
		Ingestion of fish	Adult Resident	---	---	1.10×10^{-3}	1.10×10^{-3}
		Ingestion of drinking water	Adult Resident	---	---	1.17×10^{-6}	1.17×10^{-6}
		Ingestion of soil	Child to Adult Resident	---	---	1.85×10^{-5}	1.85×10^{-5}
	Ground-water Pathway	Ingestion of drinking water	Adult Resident	---	---	4.40×10^{-2}	8.54×10^{-5}
		Dermal absorption via bathing	Adult Resident	---	---	1.10×10^{-4}	2.13×10^{-7}
		Inhalation via showering	Adult Resident	---	---	---	---
		Dermal absorption via bathing	Child Resident	---	---	2.75×10^{-4}	5.34×10^{-7}
	Total			4.49×10^{-9}	4.49×10^{-9}	4.55×10^{-2}	1.21×10^{-3}
	Human Health Screening Limits			1×10^{-6}	1×10^{-6}	1	1
Ecological	*Media*	*Concentration Under Acidic Landfill Conditions*	*Concentration Under Basic Landfill Conditions*	*Ecological Criteria[b]*	*Units*		
	Surface Water	2.2×10^{-8}	2.2×10^{-8}	2.5×10^{-4}	mg/L		
	Ground-water	8.26×10^{-4}	1.65×10^{-6}	Not applicable	mg/L		
	Ambient Air	5.22×10^{-9}	5.22×10^{-9}	Not applicable	mg/m^3		
	Soil	7.23×10^{-4}	7.23×10^{-4}	3.6×10^{-1}	mg/kg		

[a]Cancer risk and hazard quotients are the same across acidic and basic landfill conditions in the surface pathway because they are a function of the same initial total waste concentration (383 mg Cd/kg).

[b]Ecological criteria for surface water from U.S. EPA National Recommended Water Quality Criteria for Aquatic Life (Freshwater/Chronic) [12] and for soil from U.S. EPA Ecological Soil Screening Level for Cd [11].

The results of the pebble counts and gradation tests were summarized in graphical form as fragment size distributions (Figure 4). On average, approximately one-quarter of module fragments were smaller than 1 cm, the fragment size used in TCLP testing, and approximately 1% of module fragments were smaller than 0.1 mm, the fragment size used in total availability testing (Table 4). Since waste characterization leaching tests exclusively use fragment sizes smaller than 1 cm, leachate potential derived from these tests represents a small fraction of the total volume of landfilled CdTe PV module waste, and is likely to result in an overestimate of leachate Cd concentration.

Table 3. Human health and ecological screening assessment of Pb from landfill disposal of utility-scale CdTe and Si PV modules

Module Type	Media Concentration	Human Health Criteria[a]	Ecological Criteria[b]
Groundwater (mg/L)			
CdTe PV	1.70×10^{-6}	1.50×10^{-2}	Not applicable
Si PV	1.73×10^{-3}	1.50×10^{-2}	Not applicable
Surface Water (mg/L)]			
CdTe PV	8.63×10^{-10}	1.50×10^{-2}	2.50×10^{-3}
Si PV	3.30×10^{-8}	1.50×10^{-2}	2.50×10^{-3}
Ambient Air (mg/m^3)			
CdTe PV	2.05×10^{-10}	2.00×10^{-4}	Not applicable
Si PV	7.86×10^{-9}	2.00×10^{-4}	Not applicable
Soil (mg/kg)			
CdTe PV	2.83×10^{-5}	1.00×10^{2}	1.10×10^{1}
Si PV	1.09×10^{-3}	1.00×10^{2}	1.10×10^{1}

[a]Human health criteria from U.S. EPA DRAS guidance [2].

[b]Ecological criteria for surface water from U.S. EPA National Recommended Water Quality Criteria for Aquatic Life (Freshwater/Chronic) [12] and for soil from U.S. EPA Ecological Soil Screening Level for Pb[14].

DRAS = Delisting Risk Assessment Software.

Table 4. Fragment size distribution of CdTe PV modules crushed in a Municipal Solid Waste Landfill in the State of Arizona, USA

	Module #1	*Module #2*	*Module #3*	*Module #4*	*Module #5*	*Average (95% Confidence Interval)*
Proportion finer than 9.5 mm[a]	12%	27%	13%	40%	44%	27% (9-46%)
Proportion finer than 2 mm	1%	1%	0%	8%	10%	4% (0-10%)
Proportion finer than 4 mm	6%	6%	0%	19%	20%	10% (0-21%)
Proportion finer than 0.1 mm	0	1%	0%	2%	3%	1% (0-3%)

[a]9.5 mm, 2 mm, 4 mm, and 0.1 mm are waste characterization test method fragment sizes used for TCLP, STLC, Column, and Total Availability testing, respectively.

TCLP = Toxicity Characteristic Leachate Procedure.

STLC = Soluble Threshold Limit Concentration.

When sample variability is considered using 95% confidence intervals for mean fragment size, less than half of module fragments are smaller than 1 cm (Table 4) with 9-46% of fragments smaller than 9.5 mm and 0-3% smaller than 0.1 mm. These ranges again indicate that leaching potential is likely overestimated by waste characterization leaching tests which use fragment sizes smaller than 1 cm.

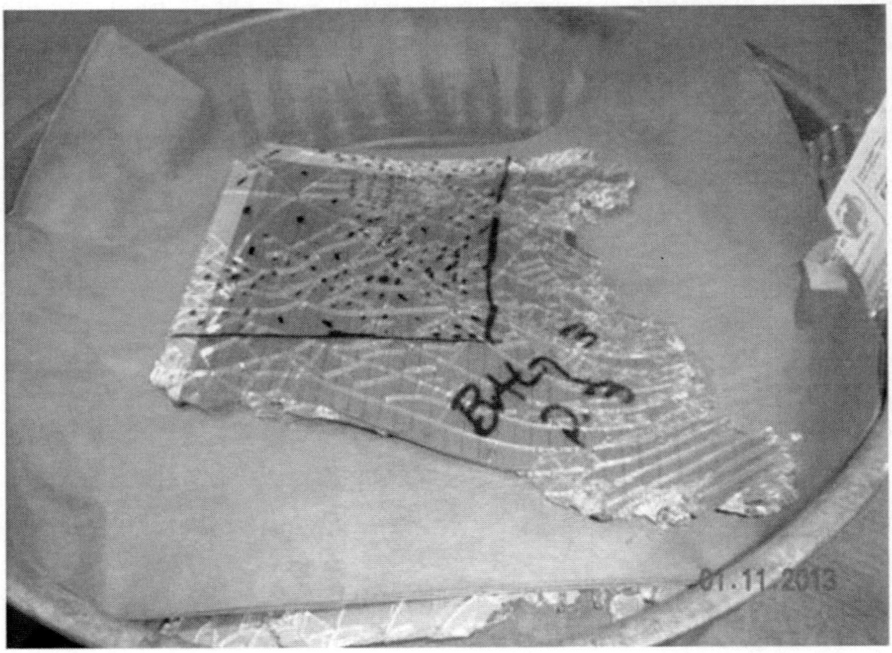

Figure 3. Compactor foot punch-out of a CdTe PV module crushed in a Municipal Solid Waste Landfill in the State of Arizona, USA.

Table 5. Waste characterization test results for CdTe PV panels crushed in a Municipal Solid Waste Landfill in the State of Arizona, USA[a]

	U.S. Regulatory Limit for Non-hazardous Waste (mg/L)	*CdTe PV Panel 2* **TCLP (mg/L)**	*CdTe PV Panel 5* **TCLP (mg/L)**	*CdTe PV Panel 2* **STLC (mg/L)**	*CdTe PV Panel 5* **STLC (mg/L)**
Lead	5	<0.1	<0.1	<0.1	<0.1
Cadmium	1	0.19	<0.1	0.57	0.91

[a]Testing of landfill fragments conducted without further reduction in fragment size to compare with standard waste characterization testing in Table 1.

TCLP = Toxicity Characteristic Leachate Procedure.

STLC = Soluble Threshold Limit Concentration.

Consistent with the relatively large fragment sizes observed, laboratory analysis of the crushed modules yielded leachate concentrations within regulatory limits for non-hazardous waste based on the TCLP waste characterization test used federally in the U.S. and the STLC test used in California [15] (Table 5). Waste characterization testing of the landfill PV fragments was commissioned by First Solar and conducted without further reduction in fragment size to compare with standard waste characterization testing (1 cm fragment size in [8]). In both cases, leachate concentrations were within regulatory limits for non-hazardous waste.

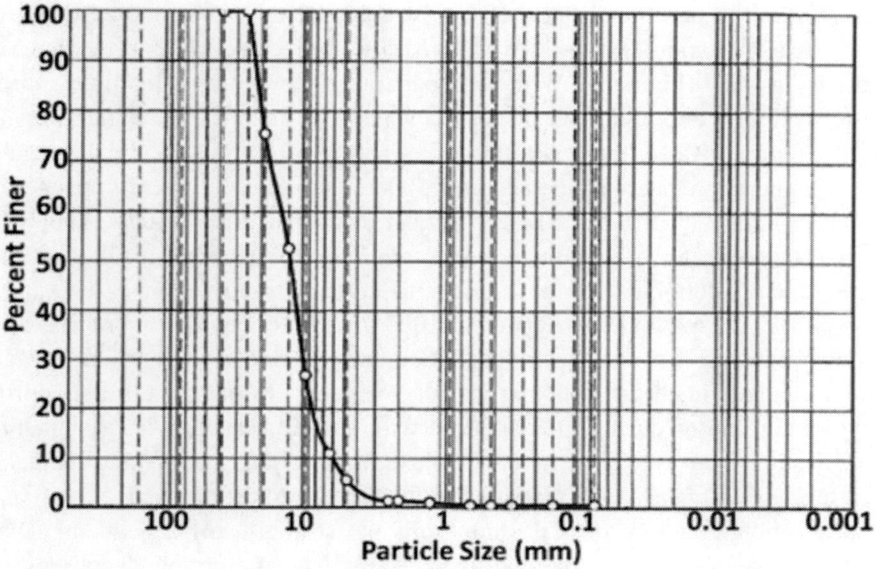

Figure 4. Fragment size distribution of CdTe PV module #2 crushed in a Municipal Solid Waste Landfill in the State of Arizona, USA.

Discussion

Previous life cycle assessment has found that emissions from CdTe PV systems are minimal in comparison with conventional electricity generation [16][17]. Previous assessments have assumed that end-of-life CdTe PV modules will be recycled. With on-site wastewater treatment and air emissions controls, emissions from CdTe PV module recycling are low, with Cd emissions to air and water of 5.89×10^{-9} and 8.92×10^{-8} kg per m^2 module, respectively [18]. Other potential end-of-life scenarios include incineration, disposal in a sanitary landfill, and disposal in a non-sanitary landfill. Incineration has not been considered here though has been evaluated by Raugei et al. [19]. With the use of engineering controls including daily cover, stormwater management, landfill liner, leachate collection, and groundwater monitoring, Cd emissions to soil (0.0001%) and to water (0.0026%) from electronics in sanitary landfills are low [20].

While recycling services and sanitary landfills are commonplace in industrialized regions such as the European Union and the United States, non-sanitary landfills are more prevalent in developing countries [21]. Therefore in this study, the third scenario (disposal in a non-sanitary landfill) has been evaluated with the U.S. EPA DRAS model. The results indicate that modeled Cd and Pb exposure from a one-time disposal of a utility-scale (25 MWac) PV project is not likely to cause adverse impacts to human health or the environment. The results presented here are generally consistent with Tvermoes et al. [6] which estimated low aggregate carcinogenic risks ($< 1\times10^{-6}$) and non-cancer hazard indices ranging from low (0.03) to slightly greater than the screening limit (1.77). The upper-bound non-cancer hazard index in Tvermoes et al. [6] resulted from a scenario that included landfilling of an 80 MWac project followed by continuous disposal of CdTe modules over the active life of the landfill

(20 years) totaling an additional 350 MWac. In contrast, this study modeled the one-time disposal of waste volume equivalent to a 25 MWac project which is a median size for a utility-scale PV project (Figure 1). While comparison of the two investigations indicates that large waste volumes and multiple disposal events can increase potential impacts from landfilled PV module waste, there is a margin of safety of 20-800 for the aggregate hazard index, 220 for the aggregate cancer risk, and 500-11,000 for ecological impacts relative to screening limits in the 25 MWacCdTe PV project evaluated in this study (Table 2). For Si PV, the margins of safety for human health impacts are a factor of 8 for impacts to groundwater and 25,000-450,000 for impacts to surface water, ambient air, and soil. In addition for Si PV, the margins of safety are 10,000-75,000 for ecological criteria (Table 3).

Another difference between this investigation and Tvermoes et al. [6] is the accounting for potential changes in the mobility of metals over the life of a landfill. Landfills move through two main phases during their lifetime with an initial anaerobic fermentation phase (acidic pH) lasting from weeks to months followed by a longer anaerobic methanogenesis phase (neutral to slightly basic pH) that can last for decades to centuries [7].

The adsorption/retention of Cd onto soils is strongly pH-dependent. During the methanogenic phase, metal ions are immobilized by the formation of metal sulfides as a result of reducing conditions. When the pH is below 6-7, cadmium is desorbed from these materials. This mechanism was tested in this investigation by using the results of different types of leachate tests as inputs for the groundwater pathway (Table 1). The aggregate hazard index under basic conditions (~0.001) was an order of magnitude lower than under acidic conditions (~0.05) (Table 2), indicating that the default DRAS assumption of acidic landfill conditions over the lifetime of the landfill can considerably overestimate potential impacts.

However, it should be noted that Kjeldsen et al. [7] postulate that after long-time periods (e.g., hundreds or even thousands of years) air may gradually intrude into the landfills, creating aerobic conditions, organic matter oxidation, increasing CO_2 concentrations and lowering of pH.Aucott [4] reported laboratory experiments where Cd previously bound as a sulfide precipitate was released to the aqueous phase where conditions were changed from slightly anaerobic to aerobic. However, Cd was subsequently removed from solution either by adsorption on iron hydroxides or by precipitation in the form of carbonates.

In addition, this investigation uses a screening level assumption that the toxicity of CdTe is equal to that of cadmium chloride and/or elemental cadmium. Based on recent toxicity testing, this assumption likely overestimates CdTe toxicity by approximately two orders of magnitude [22]. The modeling presented here also assumes that PV modules are the only source of constituents of concern in the modeled landfill and is therefore only considering incremental risk from PV disposal.

Another screening level assumption used in this analysis is related to leaching potential and fragment size. Leaching potential increases with increased surface area that results from smaller fragment size, and waste characterization tests use relatively small fragment sizes (0.1 mm to 1 cm). These fragment sizes are considerably smaller than those that would be routinely generated by typical waste handling practices related to disposal in a landfill, with on average approximately three-quarters of module fragments larger than 1 cm and approximately 99% of module fragments larger than 0.1 mm (Table 4).

Precipitation is a factor that is related to the risks associated with the disposal of PV modules in unlined landfills. Precipitation is required to produce landfill leachate and there is generally an inverse relationship between the high solar irradiation that is common to utility-scale PV project locations and the amount of annual precipitation. Many existing and planned utility-scale PV projects are located in regions that receive low amounts of precipitation [23] and therefore landfill disposal in these areas may not produce significant quantities of leachate. For example, in South Africa, landfills located in arid and semi-arid areas do not have leachate collection requirements, as water balance calculations indicate that it is unlikely that significant leachate will be generated [24].

Use of environmentally sensitive materials such as Pb, Se, Cr, and Cd compounds are commonplace in the PV industry. In addition to evaluating potential Cd emissions, this study compares DRAS-modeled Pb concentrations from landfilled utility-scale Si and CdTe PV modules, indicating no exceedances of media-specific human health or ecological screening levels (Table 3). Estimated media concentrations of Pb from Si PV in ground water, surface water, ambient air, and soil are comparable to concentrations of Cd from CdTe PV (Tables 2,3), indicating that responsible disposal is necessary for all PV technologies.

With respect to both environmental impacts and resource efficiency, the most sustainable way to manage end-of-life PV modules is recycling. The PV recycling industry is expected to scale significantly over the next 10-15 years as annual waste generated by end-of-life PV modules is predicted to increase to over 1 million tons by 2035 [25].

The recycling industry will generate jobs and recover valuable materials. For example, CdTe PV recycling technology (Figure 5) has been in commercial operation since 2005 and has already recycled over 1.5 million modules with yields of approximately 90% and 95% for glass and semiconductor recovery, respectively. End-of-life collection and recycling, including semiconductor refining, accounts for only approximately 10% of the total life cycle energy payback time of CdTe PV [18] and can result in net reductions in life cycle primary energy demand when recycling products are accounted for as credits [26].

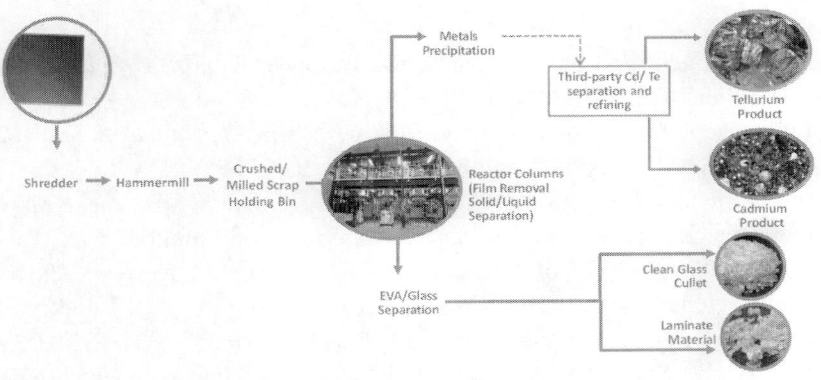

Figure 5. Schematic of CdTe PV module recycling process with 95% and 90% yields of semiconductor material and glass, respectively.

An understanding of future PV recycling can be gained by examining trends in electronic waste regulation. The EU has mandated the recycling of waste electrical and electronic equipment since 2003, and end-of-life collection and recycling of PV modules has recently been included within the scope of the recast WEEE Directive [1]. Whereas the text of the Directive depicts a mass-based recovery quota which could be obtained by recycling of PV module glass and aluminum frames only, the European Commission also clearly saw the need to develop high value recycling standards (which can include recycling of PV semiconductor material) and mandated the European Standardization Organization CENELEC to develop these standards [27]. High value recycling is necessary for maximizing resource recovery and diverting metals from disposal.

Conclusion

While recycling services and sanitary landfills are commonplace in industrialized regions such as the European Union and the United States, non-sanitary landfills are more prevalent in developing countries. With respect to the latter, based on fate and transport modeling with the U.S. EPA DRAS model, disposal of utility-scale CdTe PV modules in unlined landfills is unlikely to result in adverse health or environmental impacts. The leaching potential assumed in the fate and transport model is based on waste fragment sizes that are considerably smaller than those that would be routinely generated by typical waste handling practices in a landfill, providing an additional margin of safety. Potential impacts may be further limited by use of sanitary landfills and high value recycling, with the latter providing the important benefit of resource recovery.

Acknowledgments

We wish to thank Enamul Hoque for analysis of landfill fragment size distributions.

References

[1] European Parliament and the Council of the European Union. *Official Journal of the European Union*.2012. 24.7.2012, L 197, 38-71.
[2] U.S. Environmental Protection Agency. *RCRA Delisting Technical Support Document*; Land and Chemicals Division, Region 5: Chicago, IL, 2008.
[3] U.S. Environmental Protection Agency. (2009). *Technical Factsheet on Cadmium*. http://www.epa.gov/safewater/pdfs/factsheets/ioc/tech/cadmium.pdf
[4] Aucott, M. 2006. *The fate of heavy metals in landfills: A Review*, The New York Academy of Sciences, New York, NY.
[5] Fthenakis, V. M. *Renewable and Sustainable Energy Reviews*. 2004, 8, 303-334.
[6] Tvermoes, B.; de Sylor, M. A.; Sahmel, J.; Cyrs, W.; Paustenbach, D. 2011. *Proc. Society of Toxicology Annual Meeting*; Washington, D.C., 2011, Abstract 2909.
[7] Kjeldsen, P.; Barlaz, M.A.; Rooker, A. P.; Baun, A.; Ledin, A.; Christensen, T.H. *Critical Reviews in Environmental Science and Technology*. 2002, 32(4), 297-336.
[8] Bousselaire, J. 2013. *Analytical Report*, Test America, Irvine, CA.
[9] Arp, H. P. *Leaching from mc-Si PV module material – results from batch, column and availability tests. Comparison with thin film CdTe PV modules*; Norwegian Geotechnical Institute: Oslo, Norway, 2010.
[10] First Solar. (2012). *Projects*. http://www.firstsolar.com/en/ Projects#namerica

[11] U.S. Environmental Protection Agency. *Ecological Soil Screening Levels for Cadmium. Interim Final*; Office of Solid Waste and Emergency Response: Washington, D.C., 2005.

[12] U.S. Environmental Protection Agency. (2013). *National Recommended Water Quality Criteria. Aquatic Life Criteria Table*. http://water.epa.gov/scitech/swguidance/ standards/ criteria/current/index.cfm#altable

[13] U.S. Environmental Protection Agency. (2004). *Lead and compounds (inorganic) (CASRN 7439-92-1)*. http://www.epa.gov/iris/ subst/0277.htm#woe

[14] U.S. Environmental Protection Agency. *Ecological Soil Screening Levels for Lead, Interim Final*; Office of Solid Waste and Emergency Response: Washington, D.C., 2005.

[15] McCutcheon, C. *Analytical Report*; TestAmerica: Phoenix, AZ, 2013.

[16] Fthenakis, V.M.; Kim, H.C.; Alsema, E. *Environmental Science and Technology*. 2008, 42(6), 2168-2174.

[17] Held, M.; Ilg, R. *Progress in Photovoltaics; Research and Applications*. 2011, 19, 614-626.

[18] Sinha, P., Cossette, M.; Ménard, J.-F. *Proc. 27th EU PVSEC*; Frankfurt, Germany, 2012, 4653 – 4656.

[19] Raugei, M.; Isasa, M.; Palmer, P.F. *International Journal of Life Cycle Assessment*. 2012, 17, 192-198.

[20] European Commission. *Commission staff working paper accompanying the Proposal for a Directive of the European Parliament and of the Council on waste electrical and electronic equipment (WEEE) (recast) Impact Assessment. Annex 5*; Commission of the European Communities: Brussels, Belgium, 2008.

[21] Hoornweg, D.; Bhada-Tata, P. *What a Waste: A Global Review of Solid Waste Management*; World Bank: Washington, D.C., 2012.

[22] Kaczmar, S. 2011. *Proc. SETAC North America 32nd Annual Meeting*; Boston, MA, 2011, 226.

[23] Komoto, K.; Breyer, C.; Cunow, E.; Megherbi, K.; Faiman, D.; van der Vleuten, P. *Energy from the Desert: Very Large Scale Photovoltaic Power – State of the Art and into the Future*; Routledge: New York, NY, 2013.

[24] Johannessen L.M; Boyer, G. *Observations of Solid Waste Landfills in Developing Countries: Africa, Asia, and Latin America*; World Bank: Washington, D.C., 1999.

[25] GlobalData. *Solar Module Recycling - A Necessary Step to Maximize Environmental Benefits of Solar PV Industry*; GlobalData Research: New York, NY, 2012.

[26] Held, M. *Proc. 24th EU PVSEC*; Hamburg, Germany, 2009, 2370-2375.

[27] European Commission. *Mandate to the European Standardisation Organisations for Standardisation in the Field of Waste Electrical and Electronic Equipment (Directive 2012/19/EU (WEEE))*; Directorate – General Environment: Brussels, Belgium, 2013.

Reviewed by Vasilis Fthenakis, Columbia University and Brookhaven National Laboratory.

In: Photovoltaics
Editor: Monique A. Gill, pp. 53-71

ISBN: 978-1-63117-843-6

Chapter 3

PHOTOCURED POLYMER ELECTROLYTE MEMBRANES FOR DYE-SENSITIZED SOLAR CELLS

A. Sacco[1,*], F. Bella[1,2], S. Bianco[1] and R. Bongiovanni[2]

[1]Center for Space Human Robotics@PoliTO,
Istituto Italiano di Tecnologia, Torino, Italy
[2]Applied Science and Technology Department,
Politecnico di Torino, Torino, Italy

Abstract

The photovoltaic market is currently dominated by silicon technology, but emerging devices such as Dye-Sensitized Solar Cells (DSSCs) are drawing increasing attention, due to their easy fabrication, low cost and high conversion efficiency. However, one of the major problems limiting the long-term stability of these devices and their subsequent large-scale industrial production is the volatilization of the liquid electrolytes traditionally employed. Furthermore, a liquid also presents technological drawbacks with respect to its effective encapsulation in an electronic device.

In order to solve this problem and improve the technological perspectives of DSSCs, many studies have been focused on the preparation of quasi-solid electrolytes, in which a polymer network is exploited to effectively retain the redox mediator. In this framework, the photoinduced polymerization has attracted large attention, demonstrating to be the most effective preparation method for these polymer electrolytes. In fact, among other advantages, it is a rapid, economic, functional and environmentally friendly process, besides being easily transferable to an industrial scale.

In this chapter, the fabrication and characterization of free-standing UV-cured polymer electrolyte membranes and their integration in quasi-solid DSSCs is reported. In particular, the first part of the chapter describes the dye-sensitized solar cell working principles and the photopolymerization process. In the second part, devoted to the new electrolytes, the discussion concerns the selection of peculiar monomers useful to obtain a specific 3D network, the development of smart fillers used to selectively improve the photovoltaic parameters, and the physico-chemical correlation between charge transport properties and polymer backbone architectures.

* E-mail address: adriano.sacco@iit.it

1. Introduction

The energy issue the world faces nowadays is probably the most difficult problem to be solved from the whole human society. Distinguished studies project that the world energy consumption will raise by 56% between 2010 and 2040 [1], especially because of the boosted needs from developing countries, where demand is driven by strong, long-term economic growth. Such a terrific amount of power is currently produced mainly starting from non-regenerative sources, like fossil fuels (oil, coal, natural gas) and nuclear power, and for a minor percentage from renewable sources, like solar energy, wind power, geothermal energy, biomass and hydropower. The environmental impact of non-regenerative energy sources is extremely relevant, mainly because of the emission of greenhouse gases which causes the increase of the globally averaged temperature. Indeed it is crucial to give a strong impulse to the development of innovative technologies for the exploitation of renewable sources and, among them, solar energy has been regarded as a leading actor because of its wide availability and inexhaustibility on the human time scale.

Photovoltaics based on silicon (both mono- and poly-crystalline materials) is a well-established technology and it is currently leading the market. However, some important drawbacks, like for instance the environmental impact of the high temperature synthesis process and the inescapable high costs for the production of high purity material, certainly affect its exploitability. Moreover, the intrinsic physical characteristics of the photovoltaic process involved confine its application under direct sunlight illumination. Following these considerations, the development of very low cost photovoltaic devices, produced with cheap, widely diffused materials with a benign environmental impact, able to work properly also in low illumination conditions and with diffuse light, is a very interesting topic. One of the most promising approach for a new generation of solar devices able to fulfill these requirements is given by dye-sensitized solar cells (DSSCs), invented by O'Regan and Grätzel in 1991 [2].

The typical structure of a DSSC [3] comprises an electrode made by a nanocrystalline mesoporous film of a wide band gap semiconductor (i.e. titanium dioxide, TiO_2), a molecular sensitizer adsorbed onto the surface of the semiconductor, an electrolytic solution with a dissolved iodide/triiodide (I^-/I_3^-) redox couple, and a catalyst-covered counter-electrode. All these elements are sandwiched between two glasses covered with a thin layer of a transparent conductive oxide (usually fluorine-doped tin oxide, FTO). The impinging solar light causes the excitation of the sensitizer with the promotion of an electron towards a higher energy level and its subsequent fast injection into the conduction band of the semiconductor. The photogenerated electron travels by diffusion through the disordered network of nanoparticles until it reaches the anode of the cell, where it is collected from the FTO by the external electric circuit. Dye regeneration occurs through electron donation from the I^- reducing agent in the electrolyte, while the resulting I_3^- ions are reduced in turn at the counter-electrode, where a catalyst (usually a small amount of Pt nanoparticles or carbon-based compounds) is deposited. The structure of the device, which is basically a photoelectrochemical cell, is very simple and involves only low cost materials, with an easy production process and environmental friendliness. At present, the efficiency of DSSCs has already surpassed 11% [3], achieving the maximum value of 12.3% at the end of 2011 using a different electrolytic solution [4]. The appropriate selection and optimization of the materials constituting the cell components is essential both for obtaining high photovoltaic performance and to ensure their durability over time. Concerning the last aspect, the main criticality in DSSC technology stands in the use of an organic solvent-based liquid electrolyte [5]. A liquid redox mediator guarantees an excellent ionic transport and the formation of an interpenetrated solid-liquid junction at the semiconductor/dye/electrolyte interface, with a very efficient charge exchange. However, the volatility of the organic solvent in which the redox mediator is dissolved is the main disadvantage, since a robust and hermetic sealing of the device is particularly difficult (and it is basically not possible evolving towards a flexible architecture of the device using plastic-based substrates). Evaporation and leakages of the electrolyte, I^- weight loss and possible desorption or degradation of the dye by dissolution of water and oxygen are the main weaknesses, that could dramatically affect the long term durability of the device. For these reasons, during recent years several strategies aimed at substitute the electrolytes based on volatile organic solvents with alternative hole conducting media have been explored.

The application of electrolytes based on ionic liquids has been widely investigated, due to their good chemical and thermal stability, nonflammability, high ionic conductivity and above all their negligible vapor pressure that could guarantee a long lifetime for the device. Some valuable results were reported [6], but the high viscosity of the material and the subsequent reduced transport speed of the ionic species somehow cause mass transport problems, especially at high irradiation levels. Another promising approach is the gelation of liquid electrolyte by adding organic [7] or inorganic [8] gelling agents, but it is not expected to completely solve the problem. Inorganic (copper compounds such as CuI, CuBr, or CuSCN) or organic (spiro-OMeTAD is the most popular one) p-type semiconductors are very interesting for the fabrication of truly solid devices, but their poor transport properties strongly limit the obtainable photoconversion efficiency of the devices [9]. Thus, probably the most promising approach is the application of gel-polymers [10] or polymers [11], in which it is possible to trap a liquid electrolyte in a host matrix. Actually the combination of non-volatility, good performance, flexibility on the structure design, chemical and electrochemical stability, makes the polymer-based electrolytes the best candidates to ensure high and durable sunlight conversion efficiencies of the corresponding DSSCs [12, 13].

2. Polymer Electrolytes: Structures and Preparation Processes

With the aim of avoiding limited long-term stability, leakage of liquid electrolyte and drawbacks in hermetic sealing, quasi-solid electrolytes (QSEs) have been proposed and investigated [14]. The advantage of QSEs is that they simultaneously possess the diffusive transport property, the high ionic conductivity and the interfacial contact property of a liquid, and the cohesive property of a solid.

The category of QSEs includes three different classes of materials: liquid electrolytes gelled with inorganic nanoparticles [15], liquid electrolytes thickened with low molecular weight gelling agents [16], and polymer electrolytes [17]. The first two classes have the disadvantage that a phase separation between liquid electrolyte and gelling agent may occur over time, thus causing the decrease of the photovoltaic performances. In contrast, a polymer electrolyte is a much more homogeneous material, and exceptional long-term stability has been reported [18].

In a classical polymer electrolyte, the role of the polymer is to create a matrix able to trap and retain a fair amount of liquid electrolyte, by highly reducing the rate of leakage [19]. The cations and the anions of the redox mediator must remain free to move in the liquid phase. The preparation of these QSEs is carried out by swelling the polymeric matrix with a liquid solution containing the redox mediator: in this way, the liquid electrolyte remains trapped in the polymeric network and a stable gel-polymer is obtained. When a DSSC works, ion transport in the QSE occurs by diffusion: iodine salts (or ionic liquids) are dissociated by the organic solvent and ions are transported through the free volume of the polymeric matrix [20]. The main polymers used for the preparation of these electrolytes are poly(ethyleneoxide) [21], poly(acrylonitrile) [22], poly(oligoethyleneglycol methacrylate) [23], poly(siloxane) [24], poly(vinylidene fluoride-co-hexafluoropropylene) [25], poly(acrylamide) [26], poly(urethane) [27], poly(vinyl alcohol) [28], and poly(aniline) [29].

With the aim of making more stable and homogeneousQSEs, the crosslinking of the polymericmatrix can be performed. Crosslinked polymeric networks can be prepared either by physical crosslinking (thermo-reversible networks [30]) or by chemical crosslinking (thermosetting polymers [31]). Conventional chemical crosslinking processes occur in batch reactions, by using crosslinkers (such as 1,5-diiodopentane [32]) or performing quaternization reactions [33]. Although the products obtained can be efficient, these processes require long chemical reactions (several hours), the use of catalysts and temperature ramps, the products separation and purification. All of these steps affect both the cost and the processability of the final devices, and are definitely not suitable for the future industrialization of quasi-solid DSSCs.

As an alternative, low-cost, green and economic technique, photoinduced polymerization has been introduced as a powerful technique for the preparation of polymer electrolytes for DSSCs. Photoinduced polymerization is a reaction where initiation is triggered by a radiation [34,35]. Since most monomers do not produce initiating species with sufficiently high yields under light, a photoinitiator is usually added to the formulation. The photoinitiator is dissolved in the liquid reactive monomer; under light it produces radicals or ions, a chain reaction follows and consumes the monomer. This process leads to the formation of a polymer, which in most cases is waxy. In fact, due to the high initiation rate, the polymerization can proceed with short kinetic chains, because of enhanced bimolecular termination, thus limiting the molecular weight of the product. The photoinitiation usually requires the use of UV light, even if other radiations,including visible light, can be employed [36]. In several industrial processes, the photoinduced polymerization is performed on multifunctional monomers, thus obtaining crosslinked systems: this is the well-known UV-curing technology, originally employed in the field of coatings [37], inks [38] and adhesives [39]. Due to the fact that the polymer formation is fast (no longer than some minutes), and that the process is solvent-free and it is carried out at room temperature, in the last decades the photopolymerization has been used for many other applications, such as photolithography [40], material functionalization [41], surface modification [42], dentistry [43] and lithium battery electrolytes [44]. Due to its peculiarities, photoinitiated polymerization has recently become an elitist promising technique for the preparation of polymer electrolytes for DSSCs [13].

3. Photocurable Matrices and Membranes Fabrication

Photoinduced polymerization processes are mainly based on free radical mechanism, whose detailed kinetic description can be found elsewhere [45]; among all the monomers, acrylates are known to be the most reactive ones. When these monomers are multifunctional, the resulting UV-cured polymers exhibit a great resistance to heat, chemicals and radiations, because of their high crosslinking density. As a function of the length and of the chemical structure of the crosslinked segments, tunable physical properties can be obtained: the remarkable properties of the produced polymers justify their great commercial success [35].

Acrylates and methacrylates are the most used matrices for the preparation of UV-cured QSEs, even if α-methacryloyl-ω-methoxyocta(oxyethylene) and α-acryloyl-ω-acryloyloxyocta(oxyethylene) were usedin the very first work exploiting this polymerization technique [46]. The investigation of the effects of swelling time, crosslinking agent amount and plasticizer concentration led to a sunlight-to-electricity conversion efficiency (η) equal to 2.62%. Qin et al. fabricated QSEs by *in situ* photopolymerization of the monomer 1,6-hexanediol diacrylate (HDDA) in 1-hexyl-3-methylimidazolium iodide (HMII) [47], thus incorporating an iodide-based ionic liquid in the crosslinked polymeric matrix. The interfacial properties between the UV-cured electrolyte and the TiO_2 photoanode were improved through the use of low molecular weight non-reactive oligomers (NROs), such as poly(ethylene glycol) dimethyl ether (PEGDME), which led to the reduction of the recombination phenomena. Thus, an excellent $\eta = 5.83\%$ was achieved. Among the DSSCs assembled with photopolymerized QSEs, the best performance was obtained by Kim et al., who prepared a urethane-based UV-curable gel electrolyte, deposited on the TiO_2 electrode by means of a screen printing process [48]. The polymer precursor was an aliphatic urethane diacrylate, which reduced the rate of back electron transfer in the photoanode, leading to an efficiency equal to 6.1%.

In the framework of acrylates/methacrylates-based UV-cured membranes, a thorough investigation of the effects of the polymeric matrix on the DSSC photovoltaic performance was carried out. Many systems were prepared, each based on a difunctional oligomer and a photoinitiator (Darocur 1173, 4 wt%), and containing a monofunctional monomer (poly(ethylene glycol) methyl ether methacrylate, PEGMA, average M_n = 475); a 70:30 weight ratio between mono- and di-functional monomers was chosen. The difunctional monomers were polyethylene glycol diacrylate (PEGDA, average M_n = 258), bisphenol A propoxylate (2 PO/phenol) diacrylate (BIPDA, average M_n = 568), bisphenol A ethoxylate (2 EO/phenol) diacrylate (BIEDA, average M_n = 512) and di(ethylene glycol) dimethacrylate (DEGDMA, average M_n = 242), as reported in Figure 1. The resulting QSEs were prepared in the form of polymer electrolyte membranes (PEMs), which were swollen in a iodide/triiodide-based liquid electrolyte, remaining free-standing, non-tacky, easy to handle and extremely flexible, as shown in Figure 2.

Dynamic mechanical thermal analysis (DMTA) was performed on PEMs in order to calculate the crosslinking density (υ_e) values, which are a precious description of the polymeric networks. Interestingly, a correlation between the DSSCs photovoltaic performances and the crosslinking density values was observed. First of all, from Figure 3 it can be stated that the efficiency of the quasi-solid DSSCs depended on the amount of liquid electrolyte that was trapped in the PEMs.

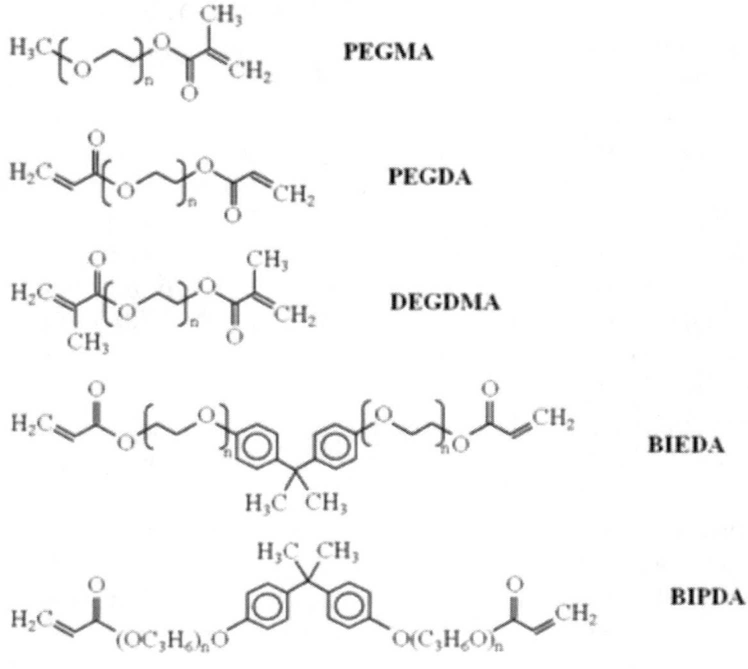

Figure 1. Structures of the reactive monomers used for the preparation of UV-cured PEMs. *n* represents the numbers of ethoxy or propoxy repeating units in the individual species and it is equal to 9 for PEGMA, 3 for PEGDA, 2 for the others.

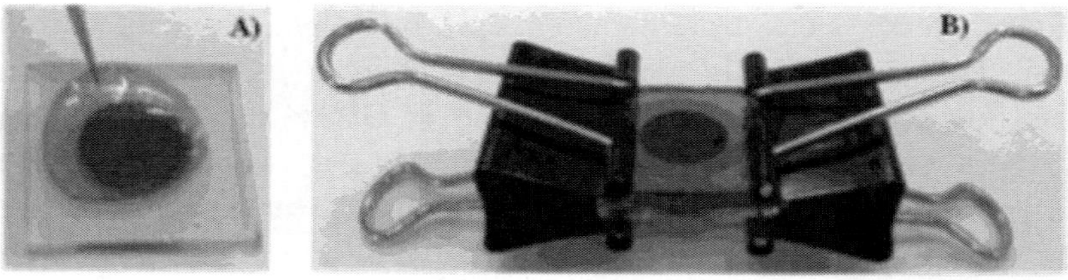

Figure 2. A) A UV-cured polymer electrolyte membrane (PEM). B) A PEM-based quasi-solid DSSC.

At high υ_e values, the polymeric network was too crosslinked to store a fair amount of liquid electrolyte. On the other hand, low υ_e values led to a very loose network, which was not so able to effectively trap the liquid electrolyte in the crosslinked network. Therefore, it is clear the importance of preparing suitable reactive mixtures with adequate monomers able to create - under UV light - a polymeric network with a crosslinking density around 1500 mol/m^3. Among the several PEMs-based DSSCs fabricated with the monomers reported in Figure 1, PEGMA:PEGDA showed the best photovoltaic performances, with an efficiency equal to 2.70%.Being a promising matrix, this was further investigated, and strategies for photovoltaic performance improvement aredetailed in Sections 4 and 5.

As previously stated, PEMs are prepared in two steps: photopolymerization of monomers and introduction of the redox couple. A particular effect can be obtained introducing the iodide salt (NaI) not only by swelling, but also as additive in the monomers reactive mixture before irradiation. It has been reported that this approach leads to a marked increase of the short circuit current density [49], and a 15-20% increase of photovoltaic performance can be obtained. This effect may be due to a synergistic interaction between the ionic transport in the liquid electrolyte trapped in PEMs and the transport which occurs through the polymer itself. While the first type of transport typically occurs by diffusion, the mechanisms of ionic conduction in polymer electrolytes can be understood considering them as extremely viscous fluids, where the local motion of the crosslinked macromolecule serves to transport ions [50]. In addition, a contribution from the ion-hopping phenomena through the rigid polymeric network should be considered as a conduction mechanism at the solid state.

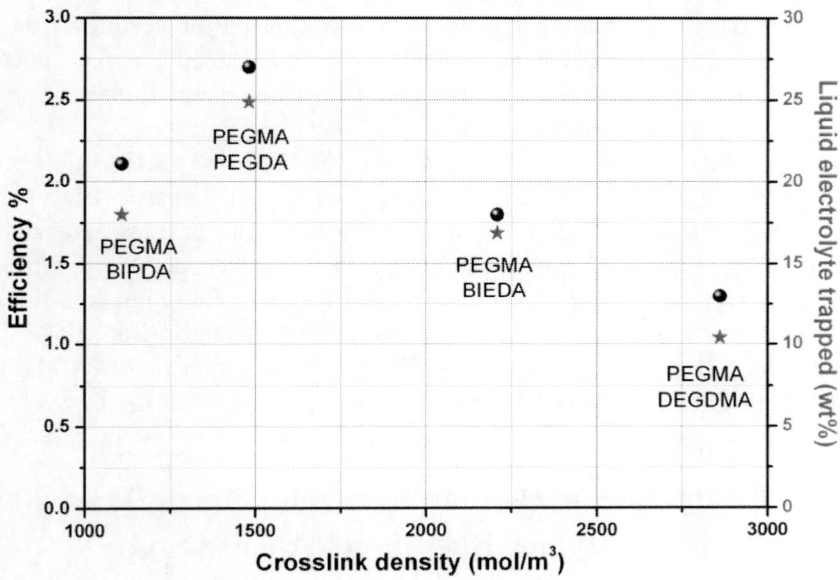

Figure 3. Light-to-electricity conversion efficiency (left axis, dark circles) and amount of liquid electrolyte trapped in the polymeric network (right axis, light stars) dependence on the crosslinking density, for four different UV-cured PEMs.

With the aim to improve the electrode/electrolyte interface and also facilitate the industrial processability, some research groups started studying the possibility of preparing the PEMs *in situ*, by UV-curing the monomers and the redox mediator between the two electrodes, when the cell is already closed. However, an awkward problem is given by the fact that iodine is a strong free-radical polymerization inhibitor. As a clever solution, Winther-Jensen et al.overcamethis problem by using TiO_2 nanoparticles as photoinitiator in a charge-transfer photopolymerization reaction [51]. Under UV irradiation, photoexcitation of TiO_2 occurred and the very reactive photogenerated holes oxidized the acrylic monomers (2-hydroxyethyl methacrylate and tetra(ethylene glycol) diacrylate), producing radical cations. These ions reacted with other monomer molecules, beginning a propagating charge-transfer polymerization process, which is not inhibited by the presence of free radical quenchers (even from the same I_2). Apart TiO_2 nanoparticles, the 1-ethyl-3-methylimidazolium thiocyanate (EMImSCN) was used: SCN^- ions form SCN^--I_2 pseudohalide complexes, which have a reduced inhibiting effect on the polymerization compared to that of iodine [52]. Even if a good photovoltaic performance was obtained (with efficiency as high as4.03%), the drawback of TiO_2-initiated charge-transfer polymerization was that 90 minutes of continuous irradiation were required, which is quite a long time and could undermine the dye photostability.

In the presence of a redox mediator different from the iodide/triiodide, it is possible to carry out the *in situ* photopolymerization. This was proposed by Bella et al., who photopolymerized a PEGMA:PEGDA reactive mixture in the presence of the pseudohalogen $SeCN^-/(SeCN)_2$ redox couple [53]. The cell was assembled by drop casting both the uncured monomers (containing the electrolyte) and the photoinitiator on the photoanode, and positioning the counter-electrode directly above it, locking with two binder clips. The crosslinking was induced by visible light to avoid the degradation of the sensitizer. A striking feature was that the *in situ* photopolymerization also allowed the sealing of the cell: once removed the binder clips after irradiation, the electrodes maintained a perfect contact between them also under mechanical stress, avoiding the use of commercial adhesive tapes, whose application at an industrial level would require machinery and time.

As regards polymeric matrices different from acrylates and methacrylates, Koh et al. proposed a QSE obtained starting from a diglycidyl ether [54]. In particular, a poly(ethylene glycol) containing terminal azide groups (PEG-N_3) was synthesized starting from poly(ethylene glycol diglycidyl ether) (PEGDGE). PEG-N_3 was complexed with iodide salts and ionic liquid, and then *in situ* photocrosslinked. However, a too high crosslinking density was obtained, which limited the energy conversion efficiency at 1.6%. Solar cell performances were then improved up to 4.2% with the addition of PEGDME as a NRO, which induced the enhancementof the ion mobilityinthepolymer electrolytes.

4. Physico-Chemical Correlation between Charge Transport Properties and Polymer Backbone: A Case Study

As stated in Section 3, among the several PEMs-based DSSCs fabricated with the monomers reported in Figure 1, PEGMA:PEGDA showed the best photovoltaic performances. In order to further enhance the photoelectrochemical properties of the DSSCs based on this class of electrolytes, the composition of the membranes was optimized by varying the PEGMA/PEGDA ratio. In particular, the different formulations under investigation are reported in Table 1.

Table 1. Composition of the PEGMA:PEGDA polymeric membranes and relative properties: thickness, gel content, temperature of 5% weight loss (T_5) and glass transition temperature (T_g)

PEGMA (wt%)	PEGDA (wt%)	Thickness(μm)	Gel content %	T_5 (°C)	T_g (°C)
0	80	100	93.2	238	67
20	60	99	93.3	232	19
40	40	102	93.6	225	12
60	20	98	94.1	187	-18
80	0	101	94.3	178	-40

As expected, the different contents of the two acrylic monomers are responsible for the final properties of the polymeric membranes. Usually a monofunctional monomer like PEGMA is able to reduce the crosslinking density of the polymeric network and to decrease the glass transition temperature (T_g); in this way, the membrane becomes more flexible and the mobility of the redox couple inside the polymeric matrix increases [55].

Real Time Fourier Transform Infrared (RT-FTIR) spectroscopy was employed to study the photopolymerization kinetics. In Figure 4 the conversion curves of the acrylate/methacrylate double bonds related to the different membranes as a function of the UV-irradiation time are reported. The typical profile of multifunctional system polymerization can be clearly observable: after an initial rising, once the polymer is formed the reaction rate decreases, while the concentration of the reactive species decreases and the viscosity increases.Once the gel point is reached, since the reactive species still present have scarce mobility, their diffusion is hindered and the reaction rate goes to zero [56]. The height of the curve at the gel point (at the plateau) represents the double bond conversion final value, while the slope of the curve at the initial stage of the UV-curing process (reported in inset of Figure 4) gives an indication of the polymerization rate.

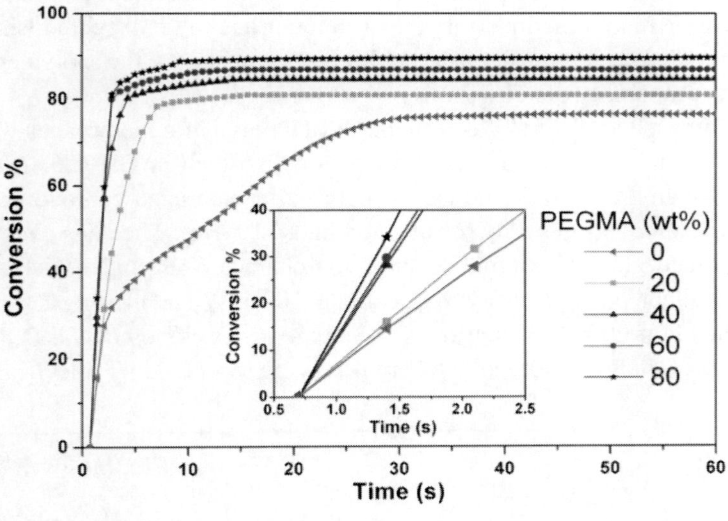

Figure 4. Acrylate/methacrylate double bonds conversion of PEGMA:PEGDA reactive mixtures evaluated throughRT-FTIR spectroscopy as a function of the UV irradiation time (the inset shows the high magnification of the curves at the initial stage of the process).

For the different membranes, final conversion values are comprised between 77% and 91% after only one minute of irradiation, meaning that the photopolymerization reaction is a very fast technique if compared with the traditional thermal processes. Both final conversion value and polymerization rate increased for larger amount of PEGMA present in reactive mixture. This fact has to be expected since a decrease in the chain propagation can occur while increasing the viscosity of the medium.

The gel content (i.e. the insoluble fraction) of the cured membranes was evaluated through the standard test method ASTM D2765-84 [53], consisting in estimating the mass reduction when the polymer is immersed in $CHCl_3$ and the non-crosslinked products are dissolved. The obtained values for the different formulations are reported in Table 1. Since all the gel content values are higher than 93%, it can be concluded that all the membranes are characterized by the formation of a nearly fully crosslinked network. In particular, the gel-content value trend is the same of the RT-FTIR measurements.

The thermal stability of the fabricated membranes was investigated through thermo-gravimetric analysis. All the measured samples are characterized by a single step of thermal degradation, occurring at temperatures larger than 350°C. In addition, the temperature T_5 at which the weight loss is 5% (usually considered the temperature at which the degradation begins) is for all samples higher than 170 °C (see Table 1), thus indicating that the thermal stability is adequate for the application of the membranes as quasi-solid electrolytes in DSSCs (the typical operating temperature is, in fact, lower than 100°C [57]). As expected, T_5 values reduce while increasing the amount of PEGMA.

In order to evaluate the viscoelastic properties of the polymeric membranes, DMTA measurements were performed. In particular, the storage (E') and loss (E'') modules were measured and the damping factor *tan* δ (calculated as E''/E' ratio) was evaluated. All the *tan* δ curves present a maximum at a proper temperature, which is usually assumed as the glass transition temperature T_g of the polymer (reported in Table 1); these peaks shifted toward lower temperature while increasing the PEGMA content inside the membrane. The reduction of the T_g for larger monofunctional monomer amount has to be expected, since the looser is the polymeric matrix, the easier will be the movements of the polymeric chains, thus diminishing the probability of vetrification.

Before being used as electrolytes for DSSCs, the polymeric membranes were swollen in a liquid solution containing the I^-/I_3^- redox mediator dissolved in CH_3CN. After this process, each UV-cured membrane was sandwiched between a platinized cathode and a nanoporous TiO_2-based photoanode (previously sensitized with N719 dye). The cells were then completed using cyanoacrylate glue as external sealant, and characterized under AM1.5G illumination (i.e. with an incident power density W_{inc} equalto 100 mW/cm^2), using a solar simulator as light source. In Figure 5 the current density-voltage (J-V) curves of the DSSCs are plotted, while the relative values of photovoltaic parameters are reported in Table 2.

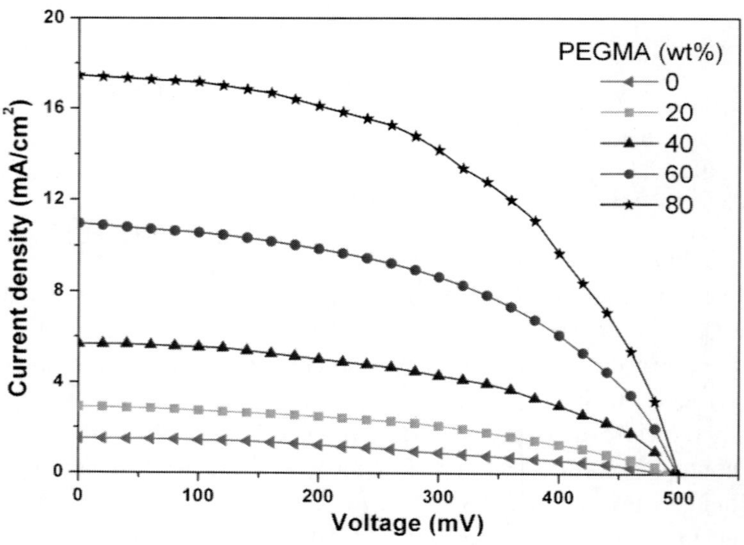

Figure 5. *J-V* curvesof PEGMA:PEGDA-based quasi-solidDSSCs.

Table 2. Photovoltaic parameters of PEGMA:PEGDA-based quasi-solid DSSCs: J_{sc} short circuit current density, V_{oc} open circuit voltage, *ff* fill factor, η efficiency

PEGMA (wt%)	PEGDA (wt%)	J_{sc} (mA/cm^2)	V_{oc} (mV)	*ff*	η (%)
0	80	1.52	489	0.46	0.36
20	60	2.91	491	0.47	0.72
40	40	5.68	495	0.49	1.42
60	20	10.96	496	0.50	2.71
80	0	17.46	499	0.52	4.41

There, a large difference can be observed regarding the photovoltaic performances of the diverse membrane-based cells. This distinction is particularly evident in the short circuit current density values J_{sc}, even if a little enhancement of the fill factor *ff* while increasing the PEGMA content can be also appreciable; on the other hand, quite no difference is present regarding the open circuit voltage V_{oc}. As a consequence of these results, a large variation of the efficiency η values (calculated through the equation 1)

$$\eta = \frac{J_{sc} V_{oc} ff}{W_{inc}} \tag{1}$$

was obtained for the different formulations of the polymeric membrane. In particular, the higher conversion efficiency (4.41%) was achieved with a polymeric membrane containing 80 wt% of PEGMA. It is worth to mention that, even if this result was obtained without employing any of the additives commonly used to improve the photovoltaic performance of DSSCs [58], the measured sunlight conversion is larger with respect to those recently reported in the literature for similar systems [51,59].

With the aim to relate the photovoltaic performances to properties of the polymeric membranes, DMTA and electrochemical impedance spectroscopy (EIS) measurements were performed.

Starting from the storage modulus curves, the crosslinking density υ_e of the polymeric network can be calculated according to the classical rubber theory equation:

$$\upsilon_e = \frac{E'}{3RT} \tag{2}$$

where E' is the elastic modulus at the rubber plateau, T is absolute temperature (above the T_g of the polymer) at which the storage modulus is evaluated, and R is the gas constant [60]. In this case, T was fixed at 390 K for all the samples. In Figure 6 the crosslinking density is reported as a function of the PEGMA content in the reactive mixture, together as the J_{sc} one. As it is clearly visible, the two quantities have opposite trends. In order to clarify this aspect, the amount of liquid electrolyte trapped in the polymeric membranes was measured (inset of Figure 6).

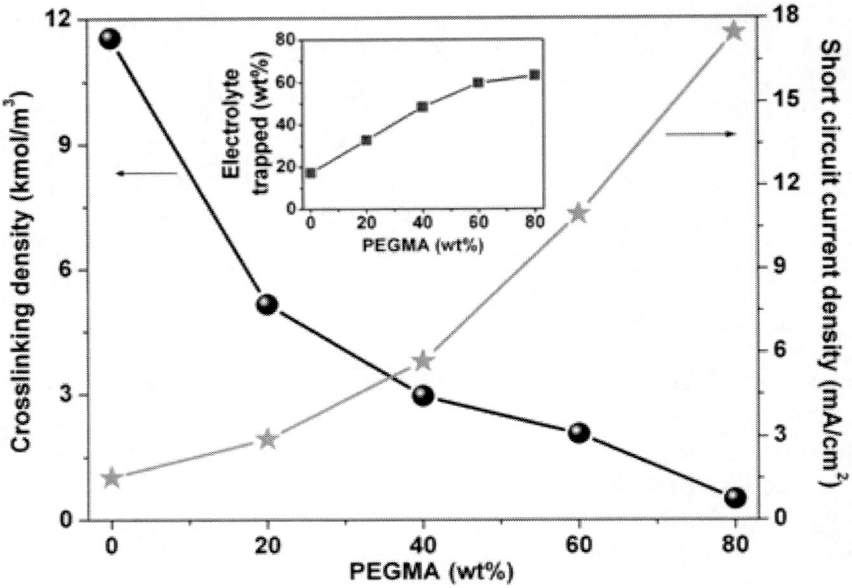

Figure 6. Crosslinking density (left axis) and short circuit current density (right axis) dependence on the amount of PEGMA content in polymerelectrolyte membranes. Inset: variation of the amount of liquid electrolyte trapped in the membranes as a function of the PEGMA content.

By comparing this curve with the $\upsilon_e$one, is it possible to conclude that a larger crosslinking density is associated with a lower amount of liquid electrolyte trapped in the polymeric matrix. In other words, by decreasing the amount of PEGMA, υ_e values increase and, obviously, the polymeric network gradually becomes too much crosslinked to effectively trap and retain a sufficient amount of liquid electrolyte. As a consequence of this fact, the short circuit current density of the quasi-solid DSSCs is reduced for lower PEGMA content, and the increase of the amount of trapped liquid electrolyte is the explanation for the enhancement of J_{sc} values [29].

Concerning the fill factor dependence on the PEGMA:PEGDA ratio, the analysis of EIS data could lead to find a relationship between the membrane composition and the total series resistance of the DSSCs (accountable for the *ff* value). To this purpose, electrochemical impedance spectroscopy measurements were carried out on symmetrical cells fabricated by sandwiching the swollen polymeric membranes between two identical platinized cathodes. The Nyquist plots of the impedance spectra related to the most performing samples (namely with 40, 60 and 80 wt% PEGMA content) are reported in Figure 7. The high frequency arc (i.e the leftmost one) is related to the charge transfer at membrane/electrode interface, and since its dimensions reduce while increasing the PEGMA:PEGDA ratio in the membranes, it is possible to conclude that the larger is the amount of PEGMA in the formulations, the lower is the charge transfer resistance. The lower frequency arc accounts for the transport of the redox mediator inside the electrolyte membrane, which is a process mainly driven by diffusion, depending on the diffusion coefficient, on the concentration of the charges and on the distance between the electrodes. The measured curves were fitted using an equivalent electrical circuit [61] in which the charge transfer impedance is modeled through the parallel $R_{ct}//C_{dl}$ and the electrolyte diffusion impedance through a Warburg element. From the fitting parameter R_{ct}, C_{dl} e ω_D (which is the diffusion characteristic frequency), the charge transfer time τ and the triiodide diffusion coefficient D can be calculated through the following formulae:

$$\tau = R_{ct} C_{dl} \quad (3)$$

$$D = \frac{\omega_D d^2}{4} \quad (4)$$

where d is the polymer thickness. These two quantities were evaluated for the different membranes. The charge transfer time dependence on the PEGMA concentration is reported in the inset of Figure 7: as it is clearly evident, the decrease of the transfer time for increasing PEGMA contents is in line with the enhancement of the *ff* values [53] (see Table 2 and inset of Figure 7). On the other hand, no dependence of the diffusion coefficient on the PEGMA:PEGDA ratio was evidenced (D values were found to be on the order of $2{\cdot}10^{-5}$ cm^2/s for all the samples): this result demonstrates that the polymer formulation has no influence on the charge diffusion inside the membranes, so the observed enhancement of the photocurrent for higher amount of PEGMA has only to be attributed to a higher amount of liquid electrolyte trapped in the polymeric matrix, as stated above.

As reported in Sections 1 and 2, one of the important advantages of the quasi-solid DSSCs with respect to their liquid counterpart is the longer lifetime due to the reduced probability of electrolyte evaporation and/or leakage.

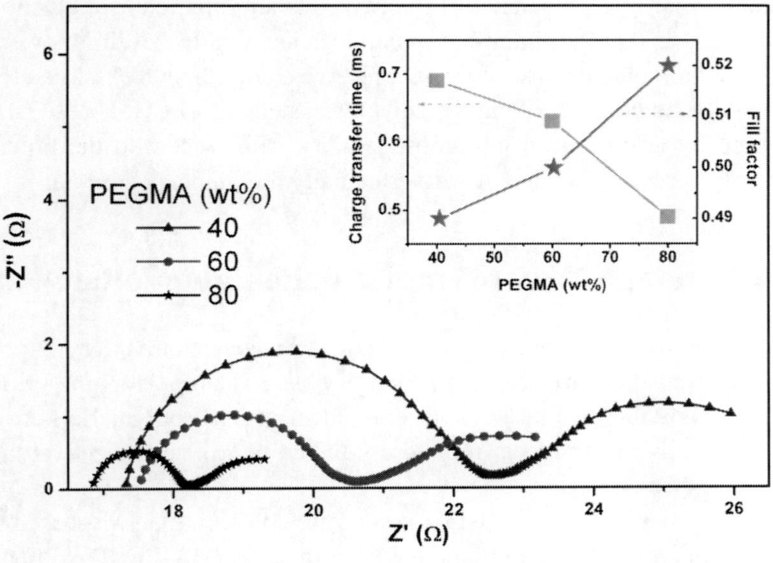

Figure 7. EIS curves of PEGMA:PEGDA-based quasi-solid DSSCs. Inset: charge transfer time (left axis) and fill factor (right axis) dependence on the amount of PEGMA content in polymer electrolyte membranes.

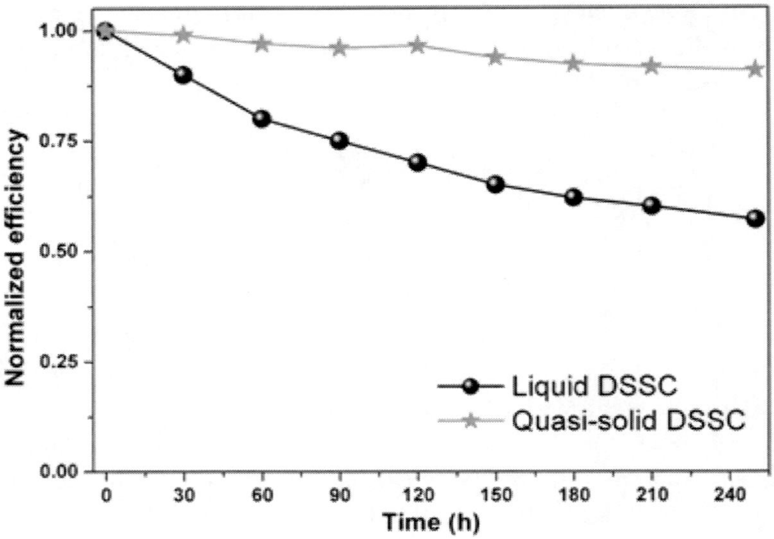

Figure 8. Variation of normalized cell efficiencies for the DSSCs fabricated with the liquid and quasi-solid (PEGMA:PEGDA 80:20) electrolytes as a function of the time.

In order to check this feature with the PEGMA:PEGDA-based cell, the most performing one (i.e. the one with PEGMA content equal to 80 wt%) was kept at the constant temperature of 60 °C for 250 h and its efficiency was measured every 30 h, according to a common procedure exploited to study the cell ageing [58]. The same procedures were replicated on a DSSC based on bare liquid electrolyte (whose initial efficiency was 6.34%) for comparison. In Figure 8 the normalized efficiencies of the two cells are reported as a function of the time. It is possible to notice that the liquid-based cell efficiency after 250 h is lower than 60% of the initial value, while the quasi-solid DSSC preserves more than 90% of its efficiency in the same period. In can be concluded that the cells fabricated with PEGMA:PEGDA membranes are characterized by an improved long-term stability with respect to the liquid-based cells, thanks to the polymeric matrix capability to effectively retain the electrolyte.

5. Filler Addition, a Way to Improve the Photovoltaic Parameters

As stated in the previous paragraphs, the monomers selection is very important in order to obtain light-cured networks able to work well as cages for the redox mediator. Furthermore, it is possible to introduce several fillers in such polymeric networks, which - after light curing - can be entrapped in the cross-linked matrix and selectively increase some photovoltaic parameters.

Recently, the preparation of a polymer composite containing an Mg-based metal-organic framework (MOF, Figure 9A) as an inorganic filler in a PEGMA-PEGDA PEM was reported [62]. Apart observing that the presence of the filler increased the thermal stability of the PEMs, it was reported the raise of V_{oc} values as a function of the Mg-MOF content. This rise was due to a variation of the conduction band edge of TiO_2 electrode, due to its coordination with the terminal carboxyl groups of the Mg-MOF particles present at the electrolyte/semiconductor interface. Efficiency values equal to 4.8% were achieved.

Furthermore, after 500 h of ageing test, the MOF-based device (Figure 9B) was able to maintain efficiency equal to the 96% of the initial value. Since the same trend was not observed for a MOF-free cell, it was stated that the particles of Mg-MOF acted as a stabilizing filler, thus protecting the UV-cured polymeric matrix from degradation under stress.

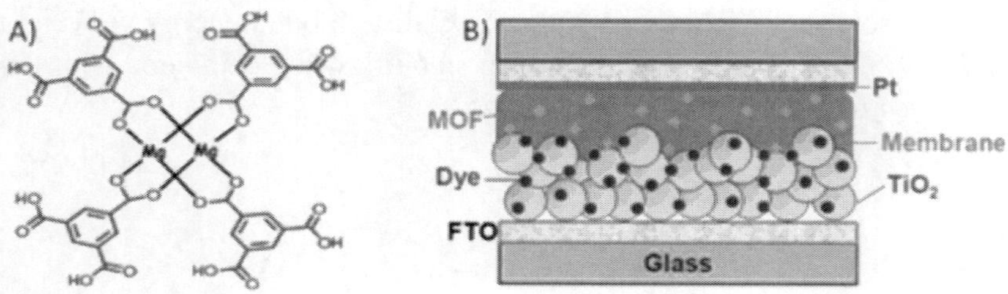

Figure 9. A) Structural formula of Mg-MOF. B) Schematic representation of a quasi-solid DSSC assembled with a Mg-MOF/PEGMA:PEGDA PEM.

With the aim to achieve at the same time the high efficiency typical of gels and the prolonged long-term stability of PEMs, a novel method of fabricating DSSCs with a double layer QSE was proposed by Han et al. [63]. A UV-cured urethane-based PEM was used as a high stability layer on the TiO_2 photoelectrode, and an additional gel electrolyte layer containing silica nanoparticles in a ionic liquid was used as a photon-to-current conversion efficiency enhancing layer. In fact, the filler created a high I^-/I_3^- concentration region around its nanoparticles, thus forming a channel useful to increase the mobility of the redox mediator in the polymeric network. An outstanding efficiency equal to 7.89% was achieved, which slightly decreased (6.82%) after two weeks of continuous light-soaking.

The introduction of nanoparticles was proposed by several research groups, exploiting the unique ability of the (photo) crosslinking process to "freeze" a polymeric system with its fillers. The effects of incorporation of aluminum nitride (AlN) in a poly(vinyidene fluoride-co-hexafluoro propylene) (PVDF-HFP)-based QSE was proposed by Huang et al. [64]. X-Ray Diffraction analysis showed a reduction of crystallinity in the polymer following the addition of AlN. As a result, the device fabricated with AlN showed J_{sc} and η values of 12.92 mA/cm^2 and 5.27%, respectively, in contrast to the corresponding values of 11.52 mA/cm^2 and 4.75% for a cell fabricated without AlN. This performance improvement was attributed to the higher apparent diffusion coefficient of I^-. Lim et al. proposed another Al-based nanosized filler (aluminum oxide, Al_2O_3). They performed the *in-situ* crosslinking of a polymer electrolyte which exhibited high ionic conductivity and favorable charge transfer reaction at the interfaces of electrodes and electrolyte, providing a remarkably high conversion efficiency equal to 6.34% [65].

Benedetti et al. proposed a crosslinked QSE composed of poly(ethyleneoxide-co-2-(2-methoxyethoxy)ethylglycidylether-co-allylglycidylether) and functionalized multi-wall carbon nanotubes (MWCNTs) [66]. In the presence of 1 wt% of MWCNT, conductivity values higher than 10^{-3} S/cm were obtained, due to the formation of a smaller concentration of polyiodide species, as confirmed by Raman spectroscopy. A further incorporation of 3 wt% of MWCNT into the QSE promoted a decrease in both J_{sc} and V_{oc}, probably due to the poorer light transmittance of the electrolyte in the visible range and to higher charge recombination losses, respectively.

Among all the fillers, also the temporary ones should be considered: they are useful when the polymer matrix is prepared, but they are subsequently eliminated when the final device is fabricated. Zhang et al. used water as filler to produce a PEG400-based PEM [67]; after water removal, a porous polymer network was obtained, which was able to trap and retain a big amount of solvated redox mediator, showing a ionic conductivity equal to $6.4 \cdot 10^{-4}$ S/cm. Quasi-solid-state solar cells fabricated with this QSE exhibited energy conversion efficiency of 6%, and also possessed improved long-term stability when compared with those based on liquid organic electrolyte.

Conclusion

Polymer electrolyte membranes were prepared from an elegant and rapid UV curing process to serve as hole conducting media in quasi-solid dye-sensitized solar cells.

In particular, four different acrylate/methacrylate-based gel–polymermembranes were prepared and activated by swelling in a solution of iodine salts. The electrochemical and photovoltaic performances of the assembled DSSCs were correlated to the crosslinking density of the polymeric network, obtained by processing the dynamo-mechanical analysis data.

Once found the monomer combination giving the most promising results (i. e. the PEGMA:PEGDA one) a study aimed to further enhance the photoelectrochemical properties of the DSSCs based on this class of electrolytes has been carried out. In particular, the composition of the membranes was optimized by varying the PEGMA: PEGDA ratio. Overall, the photovoltaic characterization of the solar harvester containing the polymer electrolyte showed valuable maximum sunlight conversion efficiency equal to 4.41%, and a significant improvement in the durability of the device with respect to the corresponding liquid counterpart was demonstrated.

The results presented here successfully demonstrate that the photoinitiated polymerization represents a powerful and rapid preparation method of polymer electrolytes for electrochemical devices, with particular focus on quasi-solid DSSCs. This technique assures the preparation of functional materials and highly efficient devices, which are characterized by improved durability upon time.

References

[1] U.S. Energy Information Administration (2013). International Energy Outlook 2013. www.eia.gov/ieo/

[2] O'Regan, B.; Gratzel, M. *Nature*, 1991, 353, 737-740.

[3] Hagfeldt, A.; Boschloo, G.; Sun, L.; Kloo, L.; Pettersson, H. *Chem. Rev.*, 2010, 110, 6595-6663.

[4] Yella, A.; Lee, H.-W.; Tsao, H. N.; Yi, C.; Chandiran, A. K.; Nazeeruddin, M. K.; Diau, E. W.-G.; Yeh, C.-Y.; Zakeeruddin, S. M.; Grätzel, M. *Science*, 2011, 334, 629-634.

[5] Yu, Z.; Vlachopoulos, N.; Gorlov, M.; Kloo, L. *Dalton Trans.*, 2011, 40, 10289-10303.

[6] Wang, P.; Zakeeruddin, S. M.; Humphry-Baker, R.; Grätzel, M. *Chem. Mater.,* 2004, 16, 2694-2696.

[7] Yu, Q.; Yu, C.; Guo, F.; Wang, J.; Jiao, S.; Gao, S.; Li, H.; Zhao, L. *Energy Environ. Sci.,* 2012, 5, 6151-6155.
[8] Wang, P.; Zakeeruddin, S. M.; Comte, P.; Exnar, I.; Grätzel, M. *J. Am. Chem. Soc.*, 2003, 125, 1166-1167.
[9] Wu, J.; Lan, Z.; Hao, S.; Li, P.; Lin, J.; Huang, M.; Fang, L.; Huang, Y. *Pure Appl. Chem.,* 2008, 80, 2241–2258.
[10] Ileperuma, O. A. *Mater. Technol.,* 2013, 28, 65-70.
[11] Nei de Freitas, J.; Nogueira, A. F.; De Paoli, M.-A. *J. Mater. Chem.,* 2009, 19, 5279-5294.
[12] Lee, J. K; Yang, M. *Mater. Sci. Eng. B.,* 2011, 176, 1142-1160.
[13] Bella, F.; Bongiovanni, R. *J. Photochem.Photobiol. C-Photochem. Rev.,* 2013, 16, 1-21.
[14] Weijia, W.; Ying, Y.; Xueyi, G.; Qinghua, T. *Chem. Bull. / Huaxue Tongbao*, 2011, 74, 144-149.
[15] Yang, Y.; Cui, J.; Yi, P.; Zheng, X.; Guo, X.; Wang, W. *J. Power Sources,* 2014, 248, 988-993.
[16] Kubo, W.; Kitamura, T.; Hanabusa, K.; Wada, Y.; Yanagida, S. *Chem. Commun.*, 2002, 374-375.
[17] Wang, G. *Acta Energ. Sol. Sin.*, 2007, 28, 517-521.
[18] Gregorio, G. L. D.; Agosta, R.; Giannuzzi, R.; Martina, F.; Marco, L. D.; Manca, M.; Gigli, G. *Chem. Commun.*, 2012, 48, 3109-3111.
[19] Wang, Y. *Sol. Energy Mater. Sol.Cells*, 2009, 93, 1167-1175.
[20] Nogueira, A. F.; Longo, C.; De Paoli, M. A. *Coord. Chem. Rev.*, 2004, 248, 1455-1468.
[21] Bella, F.; Nair, J. R.; Gerbaldi, C. *RSC Adv.,* 2013, 3, 15993-16001.
[22] Dintcheva, N. T.; Furlani, M.; Jayasundara, W. J. M. J. S. R.; Bandara, T. M. W. J.; Mellander, B. E.; La Mantia, F. P. *Rheol. Acta*, 2013, 52, 881-889.
[23] Song, C.; Zhang, L.; Wang, Y.; Yan, X.; Zhao, D. *J. Mater. Sci.,* 2013, 48, 8153-8162.
[24] Wang, F.-M.; Chu, C.-H.; Tung, Y.-L.; Hwang, B.-J.; Wang, Y.-Y.; Wan, C.-C.; Santhanam, R. *J. Solid State Electrochem.,* 2012, 16, 649-656.
[25] Peng, S.; Li, L.; Tan, H.; Srinivasan, M.; Mhaisalkar, S. G.; Ramakrishna, S.; Yan, Q. *Electrochim. Acta,* 2013, 105, 447-454.
[26] Silakul, P.; Magaraphan, R. *Adv. Mat. Res.,* 2013, 747, 325-328.
[27] Su'ait, M. S.; Ahmad, A.; Badri, K. H.; Mohamed, N. S.; Rahman, M. Y. A.; Ricardo, C. L. A.; Scardi, P. *Int. J. Hydrogen Energy*, 2014, 39, 3005-3017.
[28] Arof, A. K.; Naeem, M.; Hameed, F.; Jayasundara, W. J. M. J. S. R.; Careem, M. A.; Teo, L. P.; Buraidah, M. H. *Opt. Quantum Electron*., 2014, 46, 143-154.
[29] Tang, Z.; Liu, Q.; Tang, Q.; Wu, J.; Wang, J.; Chen, S.; Cheng, C.; Yu, H.; Lan, Z.; Lin, J.; Huang, M. *Electrochim. Acta*, 2011, 58, 52-57.
[30] Petrov, P.; Berlinova, I.; Tsvetanov, C. B.; Rosselli, S.; Schmid, A.; Zilaei, A. B.; Miteva, T.; Dürr, M.; Yasuda, A.; Nelles, G. *Macromol. Mater. Eng.*, 2008, 293, 598-604.
[31] Shim, H.; Kim, D.; Lee, C.; Kang, Y.; Suh, D. *Macromol. Res.*, 2008, 16, 424-428.
[32] Kim, Y. K.; Cho, S. E.; Hwang, W. P.; Seo, M. H.; Park, H. W.; Kim, M. R.; Lee, J. K. *Mol. Cryst. Liq. Cryst. Sci. Technol*., 2012, 566,182-192.
[33] Xiang, W.; Zhou, S.; Yin, X.; Xiao, X.; Lin, Y.; Fang, S. *Polym. Adv. Technol.*, 2009, 20, 519-523.
[34] C. Decker. *Prog. Polym Sci*., 1996, 21, 593-650.

[35] Yagci, Y.; Jockusch, S.; Turro, N. *J. Macromolecules*, 2010, 43, 6245-6260.
[36] Schroeder, W. F.; Asmussen, S. V.; Sangermano, M.; Vallo, C. I. *Polym. Int.*, 2013, 62, 1368-1376.
[37] Di Gianni, A.; Bongiovanni, R.; Turri, S.; Deflorian, F.; Malucelli, G.; Rizza, G. *J. Coat. Technol. Res.,* 2009, 6, 177-185.
[38] Guangxue, C.; Qifeng, C.; Baoling, T.; Jinglei T. *Appl. Mech. Mater.,* 2012, 182-183, 99-103.
[39] Wen, Q. Z.; Zhang, Y.; Zhu, J. H. Adv. Mat. Res. 2013, 773, 654-659.
[40] Vitale, A.; Quaglio, M.; Marasso, S. L.; Chiodoni, A.; Cocuzza, M.; Bongiovanni, R. *Langmuir.*, 2013, 29, 15711-15718.
[41] Medel, S.; Bosch, P.; Sangermano, M. *Polym. Int.*, 2014, in press, DOI:10.1002/pi.4603.
[42] Bongiovanni, R.; Marchi, S.; Zeno, E.; Pollicino, A.; Thomas, R. R. *Colloid.Surf. A-Physicochem. Eng. Asp.,* 2013, 418, 52-59.
[43] Bayou, S.; Mouzali, M.; Aloui, F.; Lecamp, L.; Lebaudy, P. *Polym. J.,* 2013, 45, 863-870.
[44] Chiappone, A.; Nair, J.; Gerbaldi, C.; Bongiovanni, R.; Zeno, E. *Cellulose,* 2013, 20, 2439-2449.
[45] Decker, C. *Macromol. Rapid Commun.,* 2002, 23, 1067-1093.
[46] Matsumoto, M.; Wada, Y.; Kitamura, T.; Shigaki, K.; Inoue, T.; Ikeda, M.; Yanagida, S. *Bull. Chem. Soc. Jpn.*, 2001, 74, 387-393.
[47] Qin, D.; Zhang, Y.; Huang, S.; Luo, Y.; Li, D.; Meng, Q. *Electrochim. Acta,* 2011, 56, 8680-8687.
[48] Lee, H.-S.; Han, C.-H.; Sung, Y.-M.; Sekhon, S. S.; Kim, K.-*J. Curr. Appl. Phys.*, 2011, 11, S158-S162.
[49] Bella, F.; Ozzello, E. D.; Bianco, S.; Bongiovanni, R. *Chem. Eng. J.,* 2013, 225, 873-879.
[50] Bruce, P.G. Solid state electrochemistry; Cambridge University Press: Cambridge, 1995; 128-143.
[51] Winther-Jensen, O.; Armel, V.; Forsyth, M.; MacFarlane, D. R. *Macromol. Rapid Commun.*, 2010, 31, 479-483.
[52] Greijer, H.; Lindgren, J.; Hagfeldt, A. *J. Phys. Chem. B.,* 2001, 105, 6314-6320.
[53] Bella, F.; Sacco, A.; Salvador, G. P.; Bianco, S.; Tresso, E.; Pirri, C. F.; Bongiovanni, R. *J. Phys. Chem.*, C 2013, 117, 20421-20430.
[54] Hwan Koh, J.; Kwan Koh, J.; Park, N.-G.; Kim, J. H. *Sol. Energy Mater. Sol. Cells,* 2010, 94, 436-441.
[55] Malucelli, G.; Gozzelino, G.; Bongiovanni, R.; Priola, A. *Polymer,* 1996, 37, 2565-2571.
[56] Wen, M.; Scriven, L. E.; McCormick, A. V. *Macromolecules,* 2001, 35, 112-120.
[57] Dai, S.; Weng, J.; Sui, Y.; Chen, S.; Xiao, S.; Huang, Y.; Kong, F.; Pan, X.; Hu, L.; Zhang, C.; Wang, K. *Inorg. Chim. Acta*, 2008, 361, 786-791.
[58] Snaith, H. J.; Schmidt-Mende, L. *Adv. Mater.*, 2007, 19, 3187-3200.
[59] Kim, S. R.; Parvez, M. K.; In, I.; Lee, H. Y.; Park, J. M. *Electrochim. Acta,* 2009, 54, 6306-6311.
[60] Hill, L. W. *Prog. Org. Coat.,* 1997, 31, 235-243.
[61] Hauch, A.; Georg, A. *Electrochim. Acta*, 2001, 46, 3457-3466.

[62] Bella, F.; Bongiovanni, R.; Kumar, R. S.; Kulandainathan, M. A.; Stephan, A. M. *J. Mater. Chem. A.,* 2013, 1, 9033-9036.
[63] Han, D. M.; Ko, K.-W.; Han, C.-H.; Kim, Y. S. *J. Mater. Chem. A.,* 2013, 1, 8529-8533.
[64] Huang, K.-C.; Chen, P.-Y.; Vittal, R.; Ho, K.-C. *Sol. Energy Mater. Sol. Cells,* 2011, 95, 1990-1995.
[65] Lim, S. J.; Kang, Y. S.; Kim, D.-W. *Electrochim. Acta.,* 2011, 56, 2031-2035.
[66] Benedetti, J. E.; Corrêa, A. A.; Carmello, M.; Almeida, L. C. P.; Gonçalves, A. S.; Nogueira, A. F. *J. Power Sources*, 2012, 208, 263-270.
[67] Zhang, X.; Wang, C.-X.; Li, F.-Y.; Xia, Y.-Y. *J. Photochem. Photobiol. A-Chem.*, 2008, 194, 31-36.

In: Photovoltaics
Editor: Monique A. Gill, pp. 73-90
ISBN: 978-1-63117-843-6

Chapter 4

ECONOMIC ANALYSIS OF SUPPORT POLICIES IN PHOTOVOLTAIC SYSTEMS: A COMPARISON BETWEEN THE TWO MAIN EUROPEAN MARKETS

L. Dusonchet and E. Telaretti [*]
DEIM: Department of Energy, Information Technology and Mathematical Models of the University of Palermo, Palermo, Italy

Abstract

The development of the photovoltaic (PV) sector in the last decade was spurred by the implementation of various support strategies aimed at reducing the gap between PV energy cost and the energy price for conventional generation. The deployment of support policies has pushed the reduction of PV energy costs, but despite this PV is still not very competitive and its development still requires adequate support mechanisms, simple grid connection procedures, and so on.

This chapter provides a technical-economic analysis of investments in PV systems, evaluating the positioning of the main European markets (Germany and Italy), with the purpose of highlighting the main differences in the implementation of support policies in the two top countries. Several case studies are examined, differentiated in relation to the rated power (from 3 kW to 1 MW) and the type of plant (ground-mounted and rooftop PV systems). The profitability indexes in terms of the payback period (PBP), net present value (NPV) and the internal rate of return (IRR) are evaluated throughout the reference period. The chapter is expected to be very useful for policy makers, energy producing industries and national Governments to estimate the profitability of PV investments in the two selected countries, predicting how the PV sector will evolve in each of the two EU states.

[*] E-mail address: telaretti@dieet.unipa.it; Fax: +39091488452. (Corresponding author: Enrico Telaretti, Technologist, Ph D)
All the authors can be found at DEIM (Department of Energy, Information Technology and Mathematical Models), University of Palermo, viale delle Scienze, Edificio 9, 90128 Palermo (Italy). (E-mail addresses: dusonchet@dieet.unipa.it, telaretti@dieet.unipa.it)

1. Introduction

The increasing concerns on climate changes, together with the growing cost of fossil fuels have pushed national Governments towards a major employment of Renewable Energy Sources (RES) in the electrical system. Especially in Europe, in the last two decades, several policies for RES support were implemented, in order to spur the growth of RES installations.

Solar energy is one of the leading technological options for the generation of electric energy in a friendly environmental way. Compared to the conventional energy sources, solar energy provides several advantages, such as [1], [2]:

- absence of greenhouse gas emissions;
- reclamation of degraded lands;
- reduced need for transmission lines;
- security and diversification of energy supply;
- greater energy independence;
- support of the deregulation of energy markets.

Solar energy can be exploited using photovoltaic (PV) or solar thermal technology, for various applications. Despite their benefits, today PV energy generates less than 1% of the total electricity supply, because of the high costs associated with this technology. In order to overcome this issue, several countries have enacted different forms of support policies in the last two decades, aiming at reducing the gap between the PV energy cost and the energy price for conventional generation. Although different forms of financing for PV have been implemented, such as capital subsidies, renewable portfolio standards (RPS), feed'in tariffs (FiTs), net-metering, VAT reduction, taxes credits, etc., in recent years mainly two types of financial incentives have survived: FiTs and capital subsidies. Nevertheless, in order to reduce the weight of support policies on national budgets, a progressive tariff decline has been planned, while maintaining an economic margin for investors.

The deployment of support policies has spurred the reduction of PV energy costs (every doubling of volume implies an approximate PV cost reduction of 20%), making possible the so-called "fuel parity" in some countries as early as 2012 and 2013 Despite this, PV is still not very competitive and its development still requires adequate support mechanisms, simple grid connection procedures and so on.

Many authors analyzed the effect of support mechanisms in the economic assessment of RES.

The effects of capital subsidies on the PV market in Europe were discussed by Gutermuth [3], Erge et al. [4], Jahn et al. [5], and Weiss et al. [6]. In [7] a short review of electricity production from RES in Slovenia is presented. The authors conclude that the most suitable ways to stimulate power production from RES are FiTs, capital subsidies and soft loans. In [8], the author develops an economical model and a computer simulation to determine the most appropriate FiT for grid-connected PV systems. The main conclusion is that, in order to have a successful PV support program, there must be a link between the financial support and the location where the PV system is installed. In [9] a parametric cost-benefit analysis is carried out, in order to investigate whether the installation of large PV plants in Cyprus is economically feasible without relevant support measures. The results indicate that the capital expenditure of a PV park is a critical parameter for the viability of the project when no FiT is available. The Greek PV market is analyzed in [10], in which authors conclude that the FiT value should be adjusted with respect to the region/area of installation, in order to avoid excessive profits for the PV owner. In [11], the authors analyze the installed solar PV and thermal collector capacities per capita for 15 EU countries, as well as their support mechanisms for PV electricity, concluding that capital subsides and incentives are vital in promoting solar thermal and PV collectors. In [12] the authors develop a linear programming model able to optimally size a grid-connected PV system for residential applications, so as to minimize the yearly customer energy cost, including PV investment cost, maintenance cost, electric utility cost and subtracting the revenue from selling the excess electricity. In [13], an analysis of PV sector in Germany, Spain and Greece is carried out, by investigating the main factors, which affect the effectiveness of PV policies. The analysis shows that, above a certain level of return, risk-related factors, such as policy instability and administrative hurdles, play a more important role in influencing investment decisions than return-related factors (such as the level of a FiT). In [14] and [15] the authors perform a comparative economic analysis of each individual European support mechanism for PV systems, based on the calculation of NPV, Discounted Cash Flows (DCF) and IRR. In [16], Focacci describes an economical evaluation of a private residential PV plant in Italy, in order to evaluate the economic feasibility according to the PV support mechanisms. In [17] Dinçer analyzes the PV electricity status and policies in Spain, Germany, USA, Japan and China, concluding that, to achieve a measurable impact on market growth and to reach a diverse cross-section of users, retail-financing terms need to be flexible. In [18], a novel methodology for calculating the economics of a residential PV system is presented. The authors use this methodology to accurately establish the economic viability of a number of PV systems in Ireland. In [19], the authors analyze the determinants of investment risk in the PV industry in Italy, through a sensitivity analysis, scenario analysis and risk analysis. They conclude by asserting that PV systems certainly will have a future in Italy, but removing the normalization limits, simplifying the procedures and providing additional supports are key factors for higher penetration of PV. Finally, in [20] the authors define a procedure for estimating the potential of PV systems in urban context in accordance with the variation of the main economic parameters and financial incentives.

The core objective of this chapter is to provide a technical-economic analysis of investments in PV systems, evaluating the positioning of the main European markets (Germany and Italy), with the purpose of evaluating the main differences in the implementation of support policies in the two top countries. Several case studies are examined, differentiated in relation to the rated power (from 3 kW to 1 MW) and the type of plant (ground-mounted and rooftop PV systems). The profitability indexes in terms of the payback period (PBP), net present value (NPV) and internal rate of return (IRR) are evaluated, throughout the reference period.

The analysis carried out in this work could help policy makers, energy producing industries and national Governments to evaluate the profitability of PV investments in the selected countries, predicting how the PV sector will evolve in each of the two EU states.

The chapter is structured as follows. After a brief description of the EU PV market, reported in Section 2, the specific situation of Germany and Italy is described in Section 3, as well as providing an historical overview of the PV promotion policies implemented in the two top EU countries. Section 4 reports the theoretical background and the operative assumptions, which form the basis of the economic evaluation. Finally, Section 5 and 6 report the discussions and the conclusions of the work, respectively.

2. The European PV Market

PV contributes, together with other RES, to achieve the targets set by the Kyoto Protocol, limiting the consumption of fossil fuels thus avoiding sanctions for signatory States in case of defaults [21].

The EU started to promote the growth of the renewable sector after the approval of Directive 2001/77/EC that provided a framework for the development of RES in Europe [22]. European support in the renewable sector has continued with the Directive 2009/28/EC (better known as the Renewable Energy Directive), establishing a new common framework for the promotion of RES energy (both electrical and thermal) [23]. The Directive set mandatory targets for the overall share of energy from RES in gross final consumption, for all EU member States, by confirming the objective of 20% as a contribution of the RES on the total European energy production in 2020, together with 20% of energy efficiency and a 20% reduction of greenhouse gas emissions, established by the European Council act 7224/1/07, 2007 [24].

In 2012, the total PV power installed in the world was 28.6 GW, bringing the cumulative PV capacity to around 96.6 GW [25] at the end of 2012. In Europe, PV systems for a total rated power equal to 17.27 GW were installed in 2012, below the 22.13 GW of 2011, bringing the PV cumulative capacity to 69.6 GW. Germany remains at the forefront, with 7.604 GW of installed PV power in 2012, followed by Italy (3.647 GW), France (1.079 GW), the U.K. (0.925 GW) and Greece (0.912 GW). Spain, despite impressive growth in past years (especially 2008), is in third place in Europe for cumulative PV capacity (4.706 GW) behind Germany (32.461 GW) and Italy (16.450 GW), but its additional PV power was very low in 2012 (223 MW).

The 96.6 GW of installed PV power all over the world corresponds to an annual PV electricity production of 115 TWh. Compared to the total world electricity consumption of 19,000 TWh, this comprises 0.6% of the total electricity demand of the world. The PV contribution to the electricity demand has surpassed 1% in several countries. Italy is in first place, with a percentage close to 7%, which corresponds to a higher percentage of 14% of peak electricity demand. Germany is in second place, with 5.6%, followed by Greece (3.5%). In Germany, in particular, the 32.46 GW of cumulated PV capacity produces, on certain days, up to 45% of the instantaneous power demand and around 14% of the electricity during peak periods [25].

Figure 1 shows the PV power installed in the two EU countries in 2012, related to the total EU PV production. Figure 2 shows the same power sharing but refers to the cumulative PV capacity at the end of 2012.

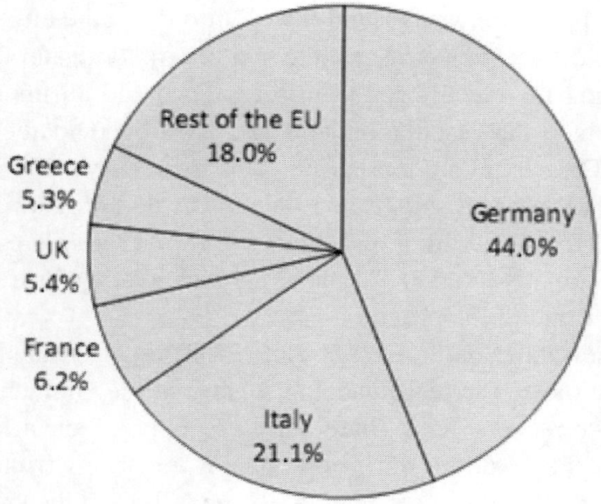

Figure 1. PV power installed in EU in 2012.

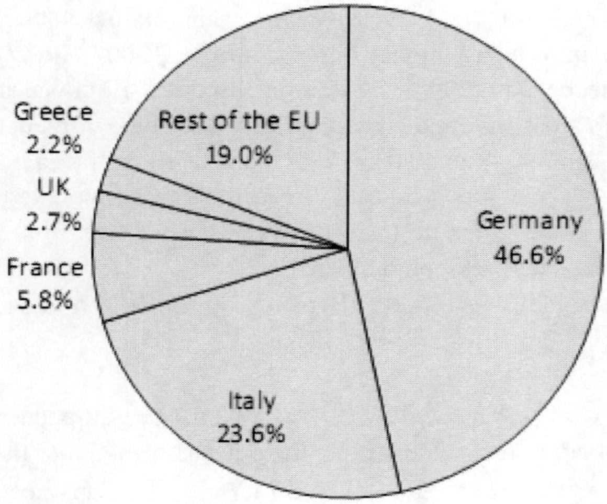

Figure 2. Cumulative PV power in the EU at the end of 2012.

3. PV Support Policies in the Selected Countries

3.1. The German Situation

Germany was one of the first countries in the world to introduce support measures for the PV sector. In 1991 the Government amended the Electricity Feed-in Act, which introduced a FiT with low tariffs for certain RES, including PV. Under this act, grid operators were obliged to pay 90% of electric retail prices for energy generated by PV plants. Furthermore, it required electricity suppliers to accept the electricity injected into the grid (with an obligation to accept the generated electricity). In the same year a capital grant promotional program was also introduced, referred to as the "Thousand Roofs Program", addressed primarily to residential customers. The program granted 70% of the investment and mounting costs for PV installations with a capacity between 1 and 5 kW. In order to be eligible for funding, the PV owner had to install 3 different electricity meters: a generation meter (metering production), a feed-in meter (metering the electricity fed into the grid) and an import meter (metering the purchased energy), since the installation operator is obliged to record the monthly meter results over 5 years. The capital grant program came to an end in 1995 and was followed by a new "Thousand Roofs Program" started in 1999. Under this program, PV installations above 1 kW were granted soft loans, with a discount of 4.5% on the interest rate. The objective of the program was to promote 100,000 PV plants with an average size of 3 kW (i.e. 300 MW) and was fulfilled at the end of the program.

In 2000, the Renewable Source Energy Act (EEG) replaced the Electricity Feed-in Act, which is currently the most effective financing scheme at the German government's disposal. The Act aims to increase the RES share of the supply of electricity by 20% by 2020, introducing a priority obligation for utilities producing electricity from RES. The FiTs are not state subsidies, but their cost is redistributed to the end user via the electricity bill. Under this new legislation, FiT values were no longer linked to retail electric prices, but fixed for 20 years. In addition, a yearly FiT reduction for newly installed PV systems was set. The EGG was subsequently amended in 2004 and 2008, increasing the RES share on electric supply to 30%by 2020 (2004), introducing higher digression rates (2004 and 2008) and introducing a flexible digression mechanism (2008). Further reform of the EEG was made in 2012 [26].

In Germany, FiTs are differentiated depending on the type of plant and the rated power of the PV installations. Building-mounted systems (such as roofs, facades or noise barriers) have higher FiT values. Lower FiT rates are established for ground-mounted systems. In Germany, PV installations with power rated up to 10 MW can benefit from FiT (PV plants higher than 10 MW was incentivized before September 2012).

In order to be eligible for support, PV plants up to 10 MW must satisfy the following conditions:

- PV systems with power rated over 100 kW must be equipped with devices that allow grid operators to reduce, at any time, the power injected into the grid;
- PV systems with power rated up to 30 kW must either meet the above-mentioned requirement or limit the power injected to the grid to 70% of the rated capacity.
- grounded-mounted systems must be built within the territorial application of a local development plan.

In 2011, Germany introduced the "Corridor" concept, in order to counteract the rapid decrease of PV costs, allowing the FiT level to decline according to market evolution. This has been achieved by splitting the FiT rate into two components: a base rate, equal to 1% and a flexible rate, dependent on the total capacity of newly installed PV systems registered in the previous 12 months. The flexible dependent rate is calculated according to a yearly government target, selected between 2.5 and 3.5 GW. The flexible digression rate is modified every month, and an update is announced every quarter, depending on the total amount of newly installed PV capacity in the past 12 months. The flexible digression rate can increase or decrease the overall digression, depending on the gap between the newly installed PV capacity and the selected yearly target, as reported in Table 1.

The latest change was announced in February 2014 by the German Federal Network Agency, with a reduction of 1% in the monthly PV FiT. For February 2014, PV installations over 1MW will receive 0.0938 €/kWh, with systems under 10 kW receiving 0.1355 €/kWh, as synthesized in Table 2.

A legal cap of 52 GW (in 2020) is established for the German FiT mechanism. When the cap is reached, FiTs will be removed on the first day of the next month.

Table 1. Total (flexible) digression rate of FiTs in Germany

Amount of deviation (MW)	
	Exceeding the target
≤ 1000	1.4% (0.4%)
> 1000	1.8% (0.8%)
≥ 2000	2.2% (1.2%)
≥ 3000	2.5% (1.5%)
≥ 4000	2.8% (1.8%)
	Below the target
≤ 500	0.75%
≤ 1000	0.5%
≤ 1500	0%
≥ 1500	0 % and the FiT is increased by 1.5%

Table 2. FiTs for electricity generated from PV systems in Germany (February, 2014)

Rated power (kW)	Installed in, at or on building or noise protection wall (€/kWh)	Freestanding facility (€/kWh)
<10	0.1355	0.0938
10.01 ÷ 40	0.1285	0.0938
40.01 ÷ 1000	0.1146	0.0938
1000.01 ÷ 10000	0.0938	0.0938

Instead of receiving a FiT, the PV owner can opt for selling the produced energy directly on the electricity market benefiting from a "market premium". The producer can switch from a FiT system to market premium and vice versa at its own convenience.

The market premium is determined each month and its value is calculated taking into account:

- the difference between the FiT rate and the average stock market price;
- a so-called management premium, calculated on top of the market premium, which covers the cost of variations in actual grid exports compared to the forecast, and stock market participation.

Like the FiT, the market premium is subject to a digression, as specified in the EEG. The management premium is also reduced every year.

In Germany, a PV owner can compensate for the energy produced by the PV system through a self-consumption scheme, thus reducing its electricity bill. The excess of PV energy injected into the grid is remunerated through a FiT system. Until 2012, German authorities incentivized self-consumption by assigning a premium above the retail electricity price. The premium was granted only for a percentage of auto-consumption higher than 30%. With the declining cost of PV technology, the bonus was removed. In the same way, a cap for FiT was established at 90% of the produced PV energy, for PV installations between 10 kW and 1 MW, in order to force self-consumption. The remaining 10%, if injected into the grid, is remunerated at a lower market price. This market integration model will take effect by January 1, 2014 and will only apply to installations erected after March 31, 2012.

On the 1st of May 2013, the government introduced a support program for promoting the growth of storage units, with the purpose of increasing self-consumption. Indeed, storage units can be used in conjunction with RES, in order to reduce the impact of renewable technologies on the electricity network (renewable energy time-shift applications) [27], [28]. Custom devices can be used to ensure a proper interconnection and a reliable control system, according to the national technical specifications [29]. The success of the incentive program is still being evaluated.

In addition to the EEG, PV receives support through long-term and low-interest loans (effective interest rate of 1.0% per year) with a fixed interest period of 5 or 10 years. In addition, facilitated connection procedures and priority connection to the grid are available for all RES plants.

3.2. The Italian Situation

In Italy, electricity generated from PV was mainly promoted through FiTs, premium tariffs and net-metering. As of July 2013, Italy stopped offering FIT remunerations because the cap of €6.7 billion had been reached. As a consequence, the FiT and the premium mechanisms cannot currently be accessed.

Since 2005 the Italian feed-in scheme, also called *ContoEnergia* (CE) has been changed five times, in an attempt to give stability and certainty to the PV market. The first CE, approved in 2005, allowed PV producers to have recourse contemporary to FiT and net-metering (also called *scambiosulposto*) for PV installations with power rated less than 20 kW (at a later stage the limit was raised to 200 kW). For power rated over 20 kW, customers could choose whether they would sell the whole PV energy to the local utility or use part of this energy for its own consumption. In the last case, the support system was composed of two terms, a FiT for all electric energy produced by the PV system, and a value for the electricity produced by the PV system, which could be used for their own consumption (with a savings in electricity cost) or partially or totally sold to the local utility. Therefore, whereas in Germany FiT was paid only for the energy effectively sold to the utility, in Italy the PV owner received a double remuneration: a FiT for all electric energy produced and a payment for the part of electric energy sold to the utility. The revenue was granted for PV systems above 1 kW, over a period of 20 years, with constant remuneration [30], [31]. FiT rates were paid by GSE (*Gestore Servizi Energetici*), the Italian Institution for the Management of Energetic Systems.

In this initial stage, the value of PV incentive depended mainly on the size of the PV system, as shown in Table 3.

Table 3. FiTs for PV systems in Italy according to the 1st CE (values referred to 2005)

Rated power (kW)	FiT (€/kWh)
$1 \leq P \leq 20$	0.445
$20 < P \leq 50$	0.460
$50 < P \leq 1000$	0.490

Table 4. FiTs for PV systems in Italy according to the 2nd CE (values referred to 2008)

Kind of installation	FiT (€/kWh)
FIPV system – 1 kW ≤ P ≤ 3 kW	0.400
FIPV system – 3 kW < P ≤ 20 kW	0.380
FIPV system – P > 20 kW	0.360
PIPV system – 1 kW ≤ P ≤ 3 kW	0.440
PIPV system– 3 kW < P ≤ 20 kW	0.420
PIPV system – P > 20 kW	0.400
BIPV system – 1 kW ≤ P ≤ 3 kW	0.490
BIPV system – 3 kW < P ≤ 20 kW	0.460
BIPV system – P > 20 kW	0.440

In 2007, a new Decree of the Ministry for Economical Development simplified the procedure to obtain the incentive and changed the FiT values distinguishing among Field Installed PV Systems (FIPV), Partially Integrated in Building (PIPV) and Building Integrated PV Systems (BIPV) [32]. The corresponding FiT values are reported in Table 4. The same Decree established that the values reported in Table 4 were decreased 2% for every year after 2008 and were increased in case of energetic certification of the building or in specific cases (PV systems realized in public school or health facilities, replacement of Eternit or asbestos, etc.).

Table 5. FiTs for PV systems in Italy according to the 3rd CE (values referred to 2011)

Kind of installation	FiT (€/kWh)
PV systems on buildings – 1 kW ≤ P ≤ 3 kW	0.402
PV systems on buildings – 3 kW < P ≤ 20 kW	0.377
PV systems on buildings – 20 kW < P ≤ 200 kW	0.358
PV systems on buildings – 200 kW < P ≤ 1000 kW	0.355
PV systems on buildings – 1000 kW < P ≤ 5000 kW	0.351
PV systems on buildings – P > 5000 kW	0.333
Other PV systems – 1 kW ≤ P ≤ 3 kW	0.362
Other PV systems – 3 kW < P ≤ 20 kW	0.339
Other PV systems – 20 kW < P ≤ 200 kW	0.321
Other PV systems – 200 kW < P ≤ 1000 kW	0.314
Other PV systems – 1000 kW < P ≤ 5000 kW	0.313
Other PV systems – P > 5000 kW	0.297

In 2010, with the approval of the third CE [33], a further distinction of PV systems was introduced: traditional PV systems, integrated PV systems with innovative features and concentrating PV systems. Traditional PV systems were themselves distinguished in PV systems on buildings and other PV systems, removing the distinction between partially and totally integrated plants of the 2nd CE. The FiT values according to the 3rd CE are reported in Table 5.

Even higher FiT rates were recognized byBIPV with innovative features (PV modules with architectural functions).

In 2011, the Legislator, in order to contain public spending, approved the 4th CE [34], establishing a FiT reduction for PV systems commissioned between June 2011 and December 2016 and introducing a simplification of PV systems in two main typologies. In addition, the 4th CE introduced the distinction between "small and great PV plants", the last being obliged to a registration in a specific register to be eligible for PV incentives.

The FiT values according to the 4th CE are reported in Table 6.

Table 6. FiTs for PV systems in Italy according to the 4th CE (values referred to 2012)

Rated power (kW)	PV systems on buildings (€/kWh)	Other PV systems (€/kWh)
$1 \leq P \leq 3$	0.387-0.274	0.344-0.240
$3 < P \leq 20$	0.356-0.247	0.319-0.219
$20 < P \leq 200$	0.338-0.233	0.306-0.206
$200 < P \leq 1000$	0.325-0.224	0.291-0.172
$1000 < P \leq 5000$	0.314-0.182	0.277-0.156
$P > 5000$	0.299-0.171	0.264-0.148

Finally, with the 5th CE, approved on August 2012, the Legislator introduced a different PV FiT scheme [35].

The Decree allows PV producers to receive support composed of two terms: a FiT for the electricity produced by the PV plant and injected into the grid and a premium for the electricity produced by the PV system and used for self-consumption. Both the FiT and the premium are reduced every quarter assuming new values reported in the Decree [35].

The producers have recourse to FiT or net-metering for PV installations with power rated less than 200 kW (the previous Decrees established that FiT and net-metering could be accumulated). For power rated over 200 kW, the PV owner can choose whether to sell all electric energy produced by the PV system to the local utility or use part of this energy for its own consumption.

The net-metering remuneration can also be cumulated with a tax deduction of 50% of the total purchase and installation PV cost (up to a maximum of € 96,000 per each residential/commercial unit). The concession allowed a reduction of 50% for 10 years of the IRPEF (regional tax on individuals) until May 31, 2013. After May 31, 2013 the concession was lowered to 36%. On the website of the Italian Authority for Electric Energy and Gas (AEEG) [36] documents are available about how to request net-metering, and on the GSE website [37] a guide on net-metering is available.

To be eligible for receiving the tariff, PV systems shall satisfy the requirements of the Italian technical standards CEI 0-16, and CEI 0-21, respectively for connection to the MV and the LV grid.

Only PV systems with power rated not below 1 kW can receive FiTs. In addition, for PV systems whose power is rated above 12 kW, the producer receives the tariff only if the plant is included in a project-funding list established by the GSE on the basis of specific requirements defined by the Decree.

The same Decree distinguishes among traditional PV systems, PV systems with innovative features and concentration PV systems. The FiT values according to the 5th CE are reported in Tables from 7 to 9.

Table 7. FiTs for traditional PV systems according to the 5th CE (April 2013)

	PV systems on building (€/kWh)		Other PV systems (€/kWh)	
Rated Power (kW)	FiT	Premium for self-consumption	FiT	Premium for self-consumption
1 ÷ 3	0.182	0.100	0.176	0.094
3.01 ÷ 20	0.171	0.089	0.165	0.083
20.01 ÷ 200	0.157	0.075	0.151	0.069
200.01 ÷ 1000	0.130	0.048	0.124	0.042
1000.01 ÷ 5000	0.118	0.036	0.113	0.031
> 5000	0.112	0.030	0.106	0.024

Table 8. FiTs for PV systems with innovative characteristics according to the 5th CE (April 2013)

Rated Power (kW)	FiT (€/kWh)	Premium for self-consumption (€/kWh)
1 ÷ 20	0.242	0.160
20.01 ÷ 200	0.231	0.149
>200	0.217	0.135

Table 9.FiTs for concentrating PV systems according to the 5th CE (April 2013)

Rated Power (kW)	FiT (€/kWh)	Premium for self-consumption (€/kWh)
1 ÷ 200	0.215	0.133
200.01 ÷ 1000	0.201	0.119
>1000	0.174	0.092

FiT values are increased by 10% if the components made inside EU member states (in particular PV modules and inverters) have an angle of incidence not below 60% of the total purchase cost (installation, transportation and design costs excluded).

As an alternative to net-metering and the FiT mechanism, PV producers can opt to sell PV energy on the free market, within the *ritiro dedicato* regulatory system. The PV energy produced is purchased by the GSE at fixed prices (GSE acts as a mediator between the producers and the electricity market). Under this mechanism, RES owners whose production is lower than 2,000 MWh can choose between a minimum tariff and the market price. The amount of energy exceeding 2000 MWh needs to be sold at the market price. The fixed guaranteed prices decrease with increasing PV capacity and are equal to €100/MWh (up to 3,750 kWh), €90/MWh (between 3,750 and 25,000 kWh) and €76.2 /MWh (between 25 and 2,000 MWh).

PV installations are also subject to a reduced VAT of 10% instead of the usual 22%.

Authorization procedures are quite simple, even though there can be differences at a territorial level. In certain cases, restrictions requiring longer procedural steps are possible.

4. Comparative Economical Analysis

4.1. Theoretical Background

The comparative economic analysis is carried out by calculating the cash flow (CF), the PBP, the NPV and the IRR for some PV test systems, for Germany and Italy, respectively.

CF depends on many factors such as the solar radiation, the FiT rate, the average electricity price, maintenance and management costs, insurance costs, etc.

The cash flow, C_t^*, for each state is calculated in the same way as in other studies by the same authors [38] ÷ [41], by summing up all the costs C_i and all the profits P_i related to the generic tth year:

$$C_t^* = \sum_i P_{i,t} - \sum_i C_{i,t} = F \cdot E_t + c_{kWh,t} \cdot E_t' - u \cdot C_0 - C_{add} \tag{1}$$

where:

- $E_t$is the energy produced by the PV system in year t;
- $F$is the FiT rate in year t;
- $E_t'$is the energy absorbed by the customer in year t (only for BIPV systems);
- $c_{kWh,t}$is the energy price per kWh in year t;
- $u$is the yearly maintenance and management cost, estimated as a percentage of the upfront cost of the PV plant, normally assumed between 0.5% and 2.4% [42]. In this study, a value of 1% is used, as in previous works [38] ÷ [41].
- $C_0$is the up-front cost of the PV installation;
- $C_{add}$is the yearly insurance cost of the PV installation.

The CF is then discounted, using the classical expression, as in prior research by the same authors:

$$C_t = C_t^* / (1 + i)^t \tag{2}$$

Where$i$ is the discount factor, assumed equal to 3%.

Equation (2) allows us to obtain the equivalent present value of the CF in year t, knowing its nominal value.

The classic methods for calculating NPV and IRR are then applied as follows:

$$NPV = \sum_{t=1}^{N} C_t^* / (1 + i)^t - C_0 \tag{3}$$

$$C_o - \sum_{t=1}^{N} C_t/(1 + IRR)^t = 0 \qquad (4)$$

where N is the lifetime of the investment.

4.2. Operative Assumptions

In order to provide an overview of different sized PV systems, the economic comparison is made considering the following five PV system categories:

1) 3 kW BIPV system;
2) 20 kW BIPV system;
3) 100 kW BIPV system;
4) 500 kW FIPV system;
5) 1000 kW FIPV system.

Table 10 shows turnkey prices per W (including VAT) for the five categories of PV plants [25], for each of the two EU countries, valid in 2012. Additional costs incurred due to the remoteness of a site or special installation requirements have not been included.

Table 10. PV installation prices, VAT included (2012)

Rated power (kW)	Germany (€/W)	Italy (€/W)
3	2.61	2.75
20	2.14	1.98
100	2.02	1.65
500	1.66	1.32
1000	1.54	1.10

In the calculations, the following hypotheses have been assumed:

- a lifespan of 25 years is assumed in all simulations;
- no capital subsidies or tax credits are considered, since they are generally applied as transitional measures and tend to have lower caps;
- in Germany, for BIPV systems (3, 20 and 100 kW), self-consumption and FiTs are both allowed;
- in Italy, only traditional PV systems have been considered;
- in Italy, since FiTs have been recently stopped, only net-metering is eligible (for BIPV systems) or the *ritiro dedicato* regulatory system (for FIPV systems);
- in Italy, when net-metering is active, the electricity produced by the PV plant is equal to the customer's energy used;
- the yearly energy produced by the PV systems is entirely sold to the utility only for FIPV systems. In Germany, for BIPV systems, 30% of the produced PV energy is transferred to the grid; the remaining 70% is self-consumed by the electricity customer;

- in Germany, after the end of the FiT granting period, the electricity produced is sold at a medium price of €50/MWh for FIPV systems. Otherwise, in Italy, the minimum prices of *ritiro dedicato* are applied;
- where net-metering or self-consumption are eligible, household electricity costs are assumed for the 3 and 20 kW PV systems, industrial electricity prices are assumed for the 100 kW PV plant.
- the mean operative efficiency of PV systems is equal to 80% [43];
- the electricity price in each state increases, every year, by 3% [14], [15].
- the yearly average energy produced by the PV systems in Germany and Italy (per kW of PV system) has the values given in Table 11, calculated using the data from RETScreen, 2013 [44];
- the average electricity cost $c_{kWh,0}$ at year 0 [45] and the inflation rates [46] in the two states have the values given in Table 12;
- the average yearly decrease in PV module production is equal to 0.8% of the total electricity generated.

Table 11. Yearly average energy produced by the PV systems

Country	Energy yearly produced (kWh/kW)
Germany	911
Italy	1290

Table 12. Average electricity cost and inflation rates in the selected EU countries, in the first quarter of 2013

Country	Household (€/kWh)	Industry (€/kWh)	Inflation rate (%)
Germany	0.292	0.143	1.1
Italy	0.229	0.168	1.3

5. Discussion

A comparison of the different ways to apply support measures for PV systems development in Germany and Italy is discussed below. Tables 13 – 14 list the PBP, the IRR and the NPV of the investments for all five PV system categories, for Germany and Italy, respectively.

Table 13. PBP, IRR and NPV for the German case

Rated power (kW)	Type of plant	Form of support	PBP (years)	IRR (%)	NPV (k€)
3	BIPV	FiT+self-consum.	15	4.81	2.15
20	BIPV	FiT+self-consum.	10.5	8.95	36.47
100	BIPV	FiT+self-consum.	19	2.19	-18.75
500	FIPV	FiT	>25	-	-
1000	FIPV	FiT	>25	-	-

Table 14. PBP, IRR and NPV for the Italian case

Rated power (kW)	Type of plant	Form of support	PBP (years)	IRR (%)	NPV (k€)
3	BIPV	Net-metering	10	9.1	7.05
20	BIPV	Net-metering	8	13.76	65.07
100	BIPV	Net-metering	9	11.58	208.56
500	FIPV	*Ritirodedicato*	16	3.94	76.26
1000	FIPV	*Ritirodedicato*	17	5.87	410.82

The value of the conventional financial indexes, listed in Tables 13-14, leads to the following fundamental considerations:

1) Ground-mounted PV plants are generally less profitable than rooftop PV systems. This is certainly due to the following reasons:
   – BIPV systems have, generally, higher FiT rates than ground-mounted systems, and the difference is not offset by the reduction in PV system costs;
   – BIPV systems are eligible for self-consumption (in Germany) or net-metering (in Italy). As a result of high retail electricity costs, the revenue arising from compensation schemes is significantly greater than those corresponding to FiT.

2) Italy has the highest profitability indexes for PV owners, mainly for BIPV systems. Indeed, for rooftop PV plants the PBP is lower than 10 years and IRR ranges between 9% and 13.7%. The high degree of profitability is certainly due to the high cost of the electricity, which makes net-metering more attractive.
3) In Italy, the only *ritiro dedicato* regulatory system is enough to make PV investment quite cost-effective, with a PBP of around 16 years and an IRR between 4% and 6%.
4) In Germany, medium and low-sided PV systems are quite profitable, thanks to self-consumption. On the other hand, IRR is negative for ground-mounted PV plants for which self-consumption is not eligible, with a PBP slightly longer than 25 years. Obviously, the presence of capital subsidies (often granted in Germany) could reverse the situation, especially for large-sized PV systems. It is useful to point out that the gap between the economic indexes for the 20 kW and the 100 kW PV systems are essentially due to the difference between the assumed electricity prices (household for the 20 kW and industry for the 100 kW PV system).

Since the highest profitability indexes arise when net-metering or self-consumption is active, it is possible to suggest that the future development of the PV sector in EU countries will be linked to the introduction of self-consumption regulatory systems even in those countries where, currently, they are not allowed.

Conclusion

This chapter has presented a technical-economic analysis of investments in PV systems, evaluating the positioning of the main European markets (Germany and Italy), with the purpose of pointing out the main differences in the implementation of support policies in the two top countries. Several case studies have been evaluated, differentiated in relation to the power rated (from 3 kW to 1 MW) and the type of plant (ground-mounted and rooftop PV system). The profitability indexes in terms of PBP, NPV and IRR have been calculated throughout the reference period. The chapter also shows that the best results are obtained when an electricity compensation scheme is active (net-metering or self-consumption), essentially because of the higher retail electricity costs compared to the FiT rates.

The results of the analysis, although in line with the experimental evidence, only cover some situations that may arise. Several factors influence the values that PBP, IRR and NPV can assume. In particular, the value of solar irradiation, which can be very different from point to point in the same country, can shift the results of the analysis towards different conclusions. Furthermore, the opportunity to take advantage of capital subsidies or tax breaks (neglected in this analysis) could significantly alter the results.

The chapter is expected to be very useful for policy makers, energy producing industries and national Legislators to evaluate the profitability of PV investments in the two selected countries, for different sizes and types of installations.

In future works, the authors will expand the analysis by considering other support measures, such as capital subsidies or tax breaks and will extend the comparison to other world countries.

References

[1] Tsoutsos, T.; Frantzeskaki, N.; Gekas, V. *Energy Policy* 2005, 33, 289-296.

[2] Solangi, K. H.; Islam, M. R.; Saidur, R.; Rahim, N. A.; Fayaz, H. *Renew and Sustain Energy Rev* 2011, 15, 2149-2163.

[3] Gutermuth, P. G. *Sol Energy* 2000, 69, 205-213.

[4] Erge, T.; Hoffmann, V. U.; Kiefer, K. *Sol Energy* 2001, 70, 479-487.

[5] Jahn, U.; Nasse, W. *Proc Third World Conf on Photovolt Energy Convers*, Osaka, Japan, 2003, pp. 2148-2151.

[6] Weiss, I.; Sprau, P.; Helm, P. *Proc Third World Conf on Photovolt Energy Convers*, Osaka, Japan, 2003, pp. 2592-2595.

[7] Stritih, U., Zupan, G., Butala, V. *Renew Sustain Energy Rev* 2007, 11, 2201-2208.

[8] Zahedi, A. *Renew Sustain Energy Rev* 2009, 13, 871-878.

[9] Poullikkas, A. *Energy Policy* 2009, 37, 3673-3680.

[10] Papadopoulos, A. M.; Karteris, M. M. *Energy Policy* 2009, 37, 1945-1952.

[11] Celik, A. N.; Muneer, T.; Clarke, P. *Renew Energy* 2009, 34, 849-856.

[12] Hongbo, R.; Weijun, G.; Yingjun, R. *Renew Energy* 2009, 34, 883-889.

[13] Lüthi S. *Sol Energy* 2010, 84, 1059-1071.

[14] Dusonchet, L., Telaretti, E.*Energy Policy* 2010, 38, 4011-4020.

[15] Dusonchet, L., Telaretti, E.*Energy Policy* 2010, 38, 3297-3308.

[16] Dinçer, F. *Renew Sustain Energy Rev* 2011, 15, 713-720.

[17] Li, Z.; Boyle, F.; Reynolds, A. *Energy* 2011, 36, 5865-5876.

[18] Focacci, A. *Renew Sustain Energy Rev* 2009, 13, 2710-2715.

[19] Cucchiella, F.; D'Adamo, I. *Renew Sustain Energy Rev* 2012, 12, 1562-1576.

[20] Orioli, A.; Di Gangi, A. *Energy* 2013, 62, 173-184.

[21] Karakosta, C.; Flouri, M.; Dimopoulou, S.; Psarras, J. *RenewSustain Energy Rev* 2012, 16, 5166-5175.

[22] Directive 2001/77/EC. European Parliament and Council of European Union. ⟨http://www.eur-lex.europa.eu/LexUriServ/LexUriServ.do?uri=CELEX:32001L0077:IT:HTML⟩; [accessed 24.01.14]

[23] Directive 2009/28/EC. European Parliament and Council of European Union. ⟨http://www.eur-lex.europa.eu/LexUriServ/LexUriServ.do?uri=OJ:L:2009:140:0016:0062:it:PDF⟩; [accessed 24.01.14].

[24] European Council, 2007. European Council act 7224/1/07, Rev. 1. ⟨http://www.consilium.europa.eu/ueDocs/cms_Data/docs/pressData/en/ec/93135.pdf⟩; [accessed 24.01.14].

[25] IEA, 2013. Trends 2013 in photovoltaic applications – Survey report of selected IEA countries between 1992 and 2012, Rep. IEA-PVPS T1-23:2013. ⟨http://www.iea-pvps.org/index.php?id=trends⟩; [accessed 24.01.14].

[26] Federal Law Gazette, 2011. Federal Law Gazette part I, n. 1634, 2011. Gesetz für den Vorrang Erneuerbarer Energien.

[27] Dusonchet, L.; Ippolito, M. G.; Telaretti, E.; Graditi, G. *Proc 9th Intern Confon the Eur Energy Market*, Florence, Italy, 2012, pp. 1-5.

[28] Dusonchet, L.; Ippolito, M. G.; Telaretti, E.; Zizzo, G.; Graditi, G. *Proc Intern Confon Power Eng Energy and Electr Drives*, Istanbul, Turkey, 2013, pp. 552-557.

[29] Ippolito, M.G.; Telaretti, E.; Zizzo, G.; Graditi, G. *Proc Fourth Intern Confon Clean Electr Power*, Alghero, Italy, 2013, pp. 262-267.

[30] D.M. 28/07/2005. Ministerial Decree of Italian Republic. ⟨http://www.autorita.energia.it/it/docs/riferimenti/decreto_050728.htm⟩; [accessed 24.01.14]

[31] D.M. 06/02/2006. Ministerial Decree of Italian Republic. ⟨http://www.conto-energia-online.it/decreto.pdf⟩; [accessed 24.01.14].

[32] D.M. 19/02/2007. Ministerial Decree of Italian Republic. ⟨http://www.autorita.energia.it/it/docs/riferimenti/decreto_070219.htm⟩; [accessed 24.01.14]

[33] D.M. 06/08/2010. Ministerial Decree of Italian Republic. ⟨http://www.camera.it/temiap/dm%206%20agosto%202010.pdf⟩; [accessed 24.01.14].

[34] D.M. 05/05/2011. Ministerial Decree of Italian Republic. ⟨http://www.gse.it/GSE_UltimiDocumenti/Conto%20energia/Fotovoltaico/02%20Normativa/DM%2005_05_2011.pdf⟩; [accessed 24.01.14].

[35] D.M. 05/07/2012. Ministerial Decree of Italian Republic. ⟨http://www.gazzettaufficiale.it/atto/serie_generale/caricaDettaglioAtto/originario?atto.dataPubblicazioneGazzetta=2012-07-10&atto.codiceRedazionale=12A07629&elenco30Giorni=false⟩; [accessed 24.01.14]

[36] AEEG, 2013. ⟨http://www.autorita.energia.it/it/index.htm⟩.

[37] GSE, 2013. ⟨http://www.gse.it/it/Pages/default.aspx⟩.

[38] Campoccia, A.; Dusonchet, L.; Telaretti, E.; Zizzo, G. *Proc. IEEE Powertech Conf.*, Lausanne, Switzerland, 2007, pp. 1981-1986.

[39] Campoccia, A.; Dusonchet, L.; Telaretti, E.; Zizzo, G. *Sol. Energy* 2009, 83, 287-297.

[40] Campoccia, A.; Dusonchet, L.; Telaretti, E.; Zizzo, G. An Analysis of Feed'in Tariffs for Solar PV in Six Representative Countries of the European Union. Sol Energy (under review).

[41] Telaretti, E.; Dusonchet, L. Comparative Economic Analysis of Support Policies for Solar PV in the most representative EU countries. Energy Policy (under review).

[42] Koner, P. K.; Dutta, V.; Chopra, K., L. *Sol Energy Matter Solar Cells* 2000, 60, 309-322.

[43] CEI 82-25, 2010. Guide for design and installation of photovoltaic (PV) systems connected to MV and LV networks. CEI – Comitato Elettrotecnico Italiano, Milano.

[44] Retscreen, 2013. ⟨http://www.retscreen.net/⟩; [accessed 24.01.14]

[45] Eurostat, 2013. *Electricity and natural gas price statistics.* ⟨http://epp.eurostat.ec.europa.eu/statistics_explained/index.php/Electricity_and_natural_gas_price_statistics⟩; [accessed 24.01.14]

[46] Eurostat, 2013. *Harmonized indices of consumer prices.* ⟨http://epp.eurostat.ec.europa.eu/cache/ITY_OFFPUB/ KS-QA-13-006/EN/KS-QA-13-006-EN.PDF⟩; [accessed 24.01.14]

In: Photovoltaics
Editor: Monique A. Gill, pp. 91-119
ISBN: 978-1-63117-843-6

Chapter 5

MODELING OF PHOTOVOLTAIC GENERATOR AS A RANDOM VARIABLE

F. J. Ruiz-Rodriguez and F. Jurado
University of Jaén, Department of Electrical Engineering EPS Linares, Jaén, Spain

Abstract

The uncertainty introduced by non-dispatchable generation, e.g. photovoltaic, affects the daily operation and planning of electric networks. To study this uncertainty, it is common to model the performance of photovoltaic units from a prediction of solar radiation. In this work a method to predict this photovoltaic generation is proposed.

The purpose of this prediction is to obtain the output electrical power from the photovoltaic generator as a random variable. This random variable can be used in studies where it is considered appropriate to take into account the uncertainties, e.g. probabilistic load flow.

The clearness index and the fraction of diffuse radiation, which are considered random variables, are the basis of the model proposed. An expression that gives the global radiation on a sloped surface is obtained, depending on the input variables that are the clearness index and the fraction of diffuse radiation. After that, another expression that gives the output electrical power from the photovoltaic generator, also as a function of the input random variables, is reached.

Cumulants method is used to solve equation for output electrical power from a photovoltaic generator. To characterize the output random variable obtained, Gram-Charlier expansion is used.

The model proposed in this work takes into account the random nature of solar radiation, dependent on the geographical location where the solar panels are installed, and the hour and day of the year in which the simulation is performing. This approach allows determining the output electrical power from a photovoltaic generator at any place and hour as random variable.

Introductión

The uncertainty introduced by non-dispatchable generation, e.g. photovoltaic, affects the daily operation and planning of electric networks, particularly in the distribution system. To study this uncertainty, it is common to model the performance of photovoltaic units from a prediction of solar radiation. The models and historical data available for photovoltaic plants will result in a forecast of electrical energy. These forecasts have implicit statistical errors, which depends on the quality of the predictions and existing models.

The importance of the predictions of the photovoltaic generation is increasing, because the penetration of photovoltaic units has increased much in last years, and it is expected that in the coming years this trend will continue. This work proposes a method to predict this photovoltaic generation [1,2].

The purpose of this prediction is to obtain the output electrical power from the photovoltaic generator as a random variable. This random variable can be used in studies where it is considered appropriate to take into account the uncertainties, e.g. probabilistic load flow.

In Section 2 of this work, some definitions and concepts related to solar radiation, which should be known before explaining the proposed model.

Section 3 focuses on explaining step by step the derivation of this model. It starts with the definitions of clearness index and fraction of diffuse radiation, which are considered random variables, whose distribution functions are also presented in this section. In this way, an expression that gives the global radiation on a sloped surface is reached, depending on the input variables that are the clearness index and the fraction of diffuse radiation.

Once this global radiation is obtained, it can be deduced another expression that shows the output electrical power from the photovoltaic generator, also as a function of the input random variables, the clearness index and the fraction of diffuse radiation.

The expression for the output electric power from the photovoltaic generator is a nonlinear equation. To solve it by the method proposed in this work, the cumulants method, it is necessary to linearize that equation. This linearization is also presented in section 3.

In section 4 the cumulants method, that is a method for solving equations in which the variables involved are random variables, is explained. The cumulants of a random variable are calculated from its statistical moments. Both the cumulants and the moments of a random variable are constants which allow knowing its properties and determining its distribution function. However, the cumulants have a number of properties that make them more useful. The calculation of the moments of a random variable is also explained in section 4, as well as relations among the moments and the cumulants.

When the method of cumulants to solve an equation with random variables is used, the cumulants of the solution random variable of the problem are obtained. To characterize this output random variable is desirable to obtain the distribution function. This is done by means of the Gram-Charlier expansion, which is explained in section 5.

Finally, in section 6, the results obtained with this method for the output electrical power from a photovoltaic generator are shown in two different situations: one in a day of winter and another in summer. The results for the probability density function and the cumulative distribution functions, as well as the probability of that the generator operates within certain power limits, are shown.

2. Some Previous Concepts about Solar Radiation

In this section some concepts which are needed for the development of the following sections are defined.

Latitude, ϕ, is the angle between the equator and a given point on Earth. If the latitude is assumed as positive in the northern hemisphere and negative in the southern hemisphere, the angular coordinates of the sun at any determined time are given by [3]:

$$\cos\theta_{zs} = \text{sen}\delta\text{sen}\phi + \cos\delta\cos\phi\cos\omega \quad (1)$$

where ω is the true solar time and δ is the solar declination, calculated as:

$$\delta = 23.45^\circ \text{sen}\left(360\frac{d_n + 284}{365}\right) \quad (2)$$

being d_n the number of the day in of the year, taking as "1" the first day of January.

Solar time, ω, is related to the official time, *TO*, that is given by the clocks, by the equation:

$$\omega = 15(TO - AO - 12) - (LL - LH) \quad (3)$$

where *LL* is the longitude of the place, *LH* is the longitude of the time zone beginning and *AO* is the official advance time.

The reference for the longitudes is the prime meridian (Greenwich) and these are taken as positives to the west and negatives to the east. In equation (3) ω, *LL* and *LH* are expressed in degrees, while *TO* and *AO* are in hours.

The angle of the sun at sunrise or sunset, ω_s, can be calculated simply by considering that the sun elevation is zero at that time, or, what is the same, making θ_{zs} =90° in equation (1). This leads to:

$$\omega_s = -\arccos(-\tan\delta\tan\phi) \quad (4)$$

In most the practical applications it is necessary to know the position of the sun with respect to the collection surface, which is defined by their slope angle, β, and orientation angle, α. The incidence angle, θ_s, of the solar radiation on the surface is given by:

$$\begin{aligned}\cos\theta_s &= \text{sen}\delta\text{sen}\phi\cos\beta - \text{sen}\delta\cos\phi\text{sen}\beta\cos\alpha + \cos\delta\cos\phi\cos\beta\cos\omega \\ &+ \cos\delta\text{sen}\phi\text{sen}\beta\cos\alpha\cos\omega + \cos\delta\text{sen}\alpha\text{sen}\omega\text{sen}\beta\end{aligned} \quad (5)$$

Despite its complicated appearance, this expression is often very easy to use. In the particular case of sloped surfaces towards the equator (α = 0), it is simplified as:

$$\cos\theta_s = \text{sen}\delta\text{sen}(\phi - \beta) + \cos\beta\cos(\phi - \beta)\cos\omega \quad (6)$$

The extraterrestrial radiation undergoes variations due to the apparent movement of the sun. These variations are perfectly predictable and can be calculated theoretically with mere geometrical considerations. For example, the extraterrestrial irradiance on a horizontal surface is:

$$B_0(0) = B_0\varepsilon_0 \cos\theta_{zs} \tag{7}$$

where B_0 is the solar constant, 1367 W/m^2, and ε_0 is the eccentricity correction factor, whose expression is:

$$\varepsilon_0 = 1 + 0.033\cos\left(360\frac{d_n}{365}\right) \tag{8}$$

The integral of the extraterrestrial irradiance along one day is:

$$B_{0,d}(0) = \frac{T}{\pi}B_0\varepsilon_0\left[-\frac{\pi}{180}\omega_s \text{sen}\phi\text{sen}\delta - \cos\phi\cos\delta\text{sen}\omega_s\right] \tag{9}$$

where T is the length of the day, i.e., 24 hours, and ω_s is expressed in degrees.

The average value for each month of this quantity is particularly relevant in the practice of engineering:

$$B_{0,dm}(0) = \frac{1}{d_{n2} - d_{n1} + 1}\sum_{d_{n1}}^{d_{n2}} B_{0,d}(0) \tag{10}$$

where d_{n1} and d_{n2} represent, respectively, the first and last days of the month considered.

2.1. Effects of Incidence Angle and the Powder

The reflectance and transmittance of the optical materials depend on the angle of incidence of the radiation. The glass which covers the photovoltaic modules is not an exception and, consequently, the efficiency of these modules is affected by the relative position between its surface with respect to the sun.

Another factor that decreases the efficiency of the photovoltaic panels, which is always present, is the powder.

A good description for these effects on the direct radiation is the expression given by Ruiz and Martin [4]:

$$FT_B(\theta_s) = 1 - \frac{\exp\left(-\frac{\cos\theta_s}{a_r}\right) - \exp\left(-\frac{1}{a_r}\right)}{1 - \exp\left(-\frac{1}{a_r}\right)} \tag{11}$$

where a_r is an adjustment parameter regarding the degree of dirt on the modules.

The following expression is for diffuse radiation:

$$FT_D(\beta) = 1 - \exp\left\{-\frac{1}{a_r}\left[c_1\left(\operatorname{sen}\beta + \frac{\pi - \frac{\beta\pi}{180} - \operatorname{sen}\beta}{1+\cos\beta}\right) + c_2\left(\operatorname{sen}\beta + \frac{\pi - \frac{\beta\pi}{180} - \operatorname{sen}\beta}{1+\cos\beta}\right)^2\right]\right\} \quad (12)$$

where c_1=4/(3π) and c_2 is linearly related to a_r. Table 1 includes the recommended values for these parameters.

Table 1. Recommended values for a_r and c_2

Dirt level	a_r	c_2
Clean	0.17	- 0.069
Low	0.20	- 0.054
Medium	0.21	- 0.049
High	0.27	-0.023

And finally, for the reflected component is:

$$FT_R(\beta) = 1 - \exp\left\{-\frac{1}{a_r}\left[c_1\left(\operatorname{sen}\beta + \frac{\frac{\beta\pi}{180} - \operatorname{sen}\beta}{1-\cos\beta}\right) + c_2\left(\operatorname{sen}\beta + \frac{\frac{\beta\pi}{180} - \operatorname{sen}\beta}{1-\cos\beta}\right)^2\right]\right\} \quad (13)$$

3. Probabilistic Photovoltaic Generator Model

The statistical properties of the total daily irradiation on a horizontal surface are important in the design and evaluation of energy conversion system performance [5]. If all days were identical, the energy delivered could be obtained by simple subtraction of the critical irradiation from the composed irradiation. Due to weather variability, the real gain of energy can be significantly higher.

Solar irradiation on a horizontal surface within the earth's atmosphere cannot be predicted accurately, because the irradiation depends on the irregular presence of clouds. Randomness and uncertainty introduced by the presence of clouds on the terrestrial radiation is usually characterized by two variables [5,6]: the daily clearness index, K_T, and the hourly diffuse fraction k_d.

Many works about characterization of the performance of solar irradiation have been published. Thus, in the works [5, 7-11] the characterization of global irradiation was studied and in [6, 12-14] the diffuse irradiation was studied. Some of them analyzed the statistical properties of the components of solar radiation, allowing to build a probabilistic model (using probability density functions, PDFs, and cumulative distribution functions, CDFs) capable to give the probability that a value of K_T or k_d, for future events, are within specified limits.

If K_T and k_d are known, it's possible to determine the random variable for global irradiance on sloped surface, $G_{g,h}(\beta,\alpha)$, as a linear combination from them.

When random variable $G_{g,h}(\beta,\alpha)$ is known, it can obtain the random variable representing the output real power of the photovoltaic generator, P_{pv}, by a linear combination of the data variables, K_T and k_d. At this point, the method of cumulants [15], for example, is used to determine the characteristics of the new random variable P_{pv} from random variables known, K_T and k_d. Then, distribution functions of output real power can be reconstructed by use Gram-Charlier expansion [16,17].

3.1. Probability Function for the Clearness Index

Initially, Liu and Jordan [8] suggested the existence of a generalized family of CDFs for K_T, only depending on itself average value, $\overline{K}_T$, despite the location and the month. From these CDFs generalized for K_T, Bent et al. [9] implemented a PDF generalized. Then, Hollands and Huget proposed the following PDF improved [5]:

$$P_K(K_T, \overline{K}_T) = C_1\left(1-\frac{K_T}{K_{Tu}}\right)\exp\left(\lambda K_T\right) \quad (14)$$

being K_{Tu} the maximum value observed of K_T (0.864). The coefficients C_1 and λ are functions of K_{Tu} and $\overline{K}_T$, and they can be determined by [5]

$$C_1 = \frac{\lambda^2 K_{Tu}}{e^{\lambda K_{Tu}} - 1 - \lambda K_{Tu}} \quad (15)$$

$$\lambda = \frac{2\Gamma - 17.519e^{-1.3113\Gamma} - 1062e^{.5.0426\Gamma}}{K_{Tu}} \quad (16)$$

where the parameter Γ is determined as:

$$\Gamma = \frac{K_{Tu}}{K_{Tu} - \overline{K}_T} \quad (17)$$

The universal nature of the CDFs of Lui and Jordan was questioned by Saunier et al. [10], since their distributions were only established from some locations. This author proposed the following PDF suitable for several climates:

$$P_K(K_T, \overline{K}_T) = C_2\left(\frac{K_T}{K_{Tu}} - \left(\frac{K_T}{K_{Tu}}\right)^2\right)\exp\left(\gamma_1\frac{K_T}{K_{Tu}}\right) \quad (18)$$

where C_2 and $\gamma_1$are functions of K_{Tu} and $\overline{K}_T$, and they are computed by the following expressions:

$$\frac{K_T}{K_{Tu}}=\frac{e^{\gamma_1}\left(\gamma_1^2-4\gamma_1+6\right)-2\gamma_1-6}{\gamma_1\left[e^{\gamma_1}\left(\gamma_1-2\right)+\left(\gamma_1+2\right)\right]} \tag{19}$$

$$C_2=\frac{\gamma_1^3}{\gamma_1\left[e^{\gamma_1}\left(\gamma_1-2\right)+\left(\gamma_1+2\right)\right]} \tag{20}$$

From equation (19), the parameter γ_1 is obtained by any iterative method, since this is an implicit equation. After obtaining this parameter, it's possible to obtain C_2 by equation (20).

3.2. Probability Function for the Fraction of Diffuse Radiation

After the first works of Page [18] and Liu and Jordan [8], many authors, e.g. [12,13, 19-24], correlated the diffuse fraction against the clearness index. These studies demonstrated that the diffuse fraction index is almost independent of the location and season. The global irradiation and its diffuse component, can be divided according to the period of integration into two groups: daily (correlations K_D-K_T) [8] and hourly (correlations k_d-k_t) [6, 12-14]. To simulate photovoltaic systems, hourly approach is preferable.

Most of the hourly correlations assume a deterministic k_d, i.e., for any given value of k_t can be only a single value of k_d, $\overline{k}_d(k_t)$. However, several authors [6,12] have pointed out that this function is not deterministic. That is, given a value of k_t, k_d can be distributed in a range of values around its mean value, $\overline{k}_d$. Strictly, k_d must be considered as an independent random variable.

Based on this approach, Hollands et al. [6] identified a general purpose expression containing the PDF for k_d:

$$P_k(k_{d,} k_t)=C_3\left(k_d-k_{dl}\right)\left(1-k_d\right)\mathrm{e}^{\lambda_1 k_d} \tag{21}$$

where C_3, $\lambda_1$andk_{dl} are functions of $\overline{k}_d$.

The expression for P_k (k_d, k_t) is independent of the relationship $\overline{k}_d(k_t)$. Any appropriate relationship for $\overline{k}_d(k_t)$ can be used [6, 12-14]. Therefore, the dependence on latitude and season of $\overline{k}_d(k_t)$ doesn't have relation to the possible universality of the equation (21).

The following will be explained the process which is followed for determining the parameters C_3, λ_1yk_{dl} in equation (21).

First, the clearness index, k_t, must be determined. This index is calculated from its namesake for daily intervals, K_T, according to the expression [5]:

$$k_t = K_T\left(a + b\cos\omega\right) \tag{22}$$

being a and b as defined in equations (33) and ω solar time, as defined in (3).

When clearness index, k_t,is known, the average value of the hourly diffuse fraction, $\overline{k}_d$, can be determined by the following correlations [13]:

$$\begin{aligned}
&\overline{k}_d = 1 - 0.09k_t \qquad \text{for} \quad k_t \le 0.22\\
&\overline{k}_d = 0.9511 - 0.16k_t + 4.388k_t^2 - 16.638k_t^3 + 12.336k_t^4\\
&\qquad \text{for} \;\; 0.22 < k_t \le 0.80\\
&\overline{k}_d = 0.165 \qquad \text{for} \quad k_t > 0.80
\end{aligned} \tag{23}$$

Through this value of average hourly diffuse fraction, $\overline{k}_d$, the parameter k_{dl} is computed by the expression [6]:

$$k_{dl} = 4.8627\overline{k}_d^4 - 5.1741\overline{k}_d^3 + 1.38\overline{k}_d - 0.1331 \tag{24}$$

Now, another parameter, ϑ, which depends on k_{dl} and $\overline{k}_d$ is calculated [6]:

$$\vartheta = \frac{\overline{k}_d - k_{dl}}{1 - k_{dl}} \tag{25}$$

When ϑ is known, the next parameter, χ, can be determined through the following correlations: [6]

$$\begin{aligned}
&\chi = \frac{-\left(1 + \vartheta + \sqrt{\vartheta^2 - 4\vartheta + 1}\right)}{\vartheta} \quad \text{for} \;\; 0 \le \vartheta \le 0.25\\
&\chi = 4.18\,\text{senh}\left[4.72\left(\vartheta - 0.5\right)\right] \quad \text{for} \;\; 0.25 \le \vartheta \le 0.75\\
&\chi = \frac{\left(1 - \vartheta + \sqrt{\vartheta^2 + 2\vartheta - 2}\right)}{1 - \vartheta} \quad \text{for} \;\; 0.75 \le \vartheta \le 1
\end{aligned} \tag{26}$$

Then, the parameter λ_1 in equation (21) is calculated by the following expression [6]:

$$\lambda_1 = \frac{\chi}{1 - k_{dl}} \tag{27}$$

And finally, the parameter C_3 is determined [6]:

$$C_3 = \frac{1}{(1-k_{dl})\left[2+\chi+(\chi-2)e^{\chi}\right]\chi^{-3}e^{\lambda_1 k_{dl}}} \tag{28}$$

Figure 1 shows a summary about the process to calculate the parameters for the probability density function of the hourly diffuse fraction.

$$k_t = K_T\left(a + b\cos\omega\right)$$

$$\begin{aligned}
\bar{k}_d &= 1-0.09k_t && \text{for } k_t \le 0.22\\
\bar{k}_d &= 0.9511-0.16k_t+4.388k_t^2-16.638k_t^3+12.336k_t^4 && \text{for } 0.22 < k_t \le 0.80\\
\bar{k}_d &= 0.165 && \text{for } k_t > 0.80
\end{aligned}$$

$$k_{dl} = 4.8627\bar{k}_d^4 - 5.1741\bar{k}_d^3 + 1.38\bar{k}_d - 0.1331$$

$$\vartheta = \frac{\bar{k}_d - k_{dl}}{1-k_{dl}}$$

$$\begin{aligned}
\chi &= \frac{-\left(1+\vartheta+\sqrt{\vartheta^2-4\vartheta+1}\right)}{\vartheta} && \text{for } 0 \le \vartheta \le 0.25\\
\chi &= 4.18\,\text{senh}\left[4.72(\vartheta-0.5)\right] && \text{for } 0.25 \le \vartheta \le 0.75\\
\chi &= \frac{\left(1-\vartheta+\sqrt{\vartheta^2+2\vartheta-2}\right)}{1-\vartheta} && \text{for } 0.75 \le \vartheta \le 1
\end{aligned}$$

$$\lambda_1 = \frac{\chi}{1-k_{dl}}$$

$$C_3 = \frac{1}{(1-k_{dl})\left[2+\chi+(\chi-2)e^{\chi}\right]\chi^{-3}e^{\lambda_1 k_{dl}}}$$

Figure 1. Calculation of the parameters of the PDF for k_d.

3.3. Global Irradiance on Horizontal Surface

To obtain the global irradiance on a horizontal surface, it will be based on the definition of clearness index [8]:

$$K_T = \frac{H_{g,dm}(0)}{B_{0,dm}(0)} \tag{29}$$

where $H_{g,dm}(0)$ is the global daily average irradiance for each month on a horizontal surface. This data is tabulated for each geographical location. $B_{0,dm}(0)$ is the extraterrestrial daily average irradiance for each month on a horizontal surface. Its calculation is explained in section 2.

The clearness index is considered a random variable: equations (14) and (18). If this randomness is taken in account, it's possible to get the global daily average irradiation for each month as a random variable, $H_{g,dm}(0)$, from equation (29):

$$H_{g,dm}(0) = B_{0,dm}(0) \cdot K_T \tag{30}$$

This equation is for daily average values for each month. To calculate the average values for intervals of one hour, i, the following relation [20] is applied:

$$H_{g,i}(0) = r_g \cdot H_{g,dm}(0) \tag{31}$$

The factor r_g is computed as:

$$r_g = \frac{180^\circ}{T}(a + b \cdot \cos\omega)\frac{\cos\omega - \cos\omega_s}{\omega_s \cos\omega_s - \operatorname{sen}\omega_s} \tag{32}$$

where ω is the solar time, calculated at the midpoint of the interval considered. ω_s is the angle of sunrise and sunset. T is the time considered within a day. For one hour periods its value is 24 and for 15 minutes periods its value is 96. In section 2 is explained how to calculate the parameters ω, and ω_s. The parameters a and $b$are computed as follow [20]:

$$\begin{aligned} a &= 0.409 - 0.5016 \cdot sen(\omega_s + 60^\circ) \\ b &= 0.6609 + 0.4767 \cdot sen(\omega_s + 60^\circ) \end{aligned} \tag{33}$$

Now it is necessary to determine the values from irradiation (energy) to irradiance (power), by the flowing expression [20]:

$$G_{g,i}(0) = N_h \cdot H_{g,i}(0) \tag{34}$$

being N_h the number of time intervals considered within an hour. For one hour intervals its value is 1, whereas for intervals of 15 min, for example, its value is 4. Performing operations, the final equation for the global irradiance on a horizontal surface is found:

$$G_{g,i}(0) = N_h \cdot r_g \cdot B_{0,dm}(0) \cdot K_T \tag{35}$$

3.4. Diffuse Irradiance on Horizontal Surface

The hourly diffuse fraction is defined as [6]

$$k_d = \frac{H_{D,h}(0)}{H_{g,h}(0)} \tag{36}$$

where $H_{D,h}(0)$ is the hourly diffuse irradiation a month on horizontal surface, and $H_{g,h}(0)$ is the hourly global irradiation for each month on horizontal surface.

The hourly diffuse fraction is a random variable, as shown in equation (21)(21). Considering this, it's possible to obtain the hourly diffuse irradiation as a random variable from equation (36):

$$H_{D,h}(0) = H_{g,h}(0) \cdot k_d \tag{37}$$

From equation (31), it can be deduced the following relationship:

$$H_{g,h}(0) = r_g \cdot H_{g,dm}(0) \tag{38}$$

being r_g the factor defined by equation (32). As in equation (38) this factor is always to T = 24, it will be called r_{g24}.

If equation (38) is replaced into (37), the following is obtained:

$$H_{D,h}(0) = r_{g24} \cdot H_{g,dm}(0) \cdot k_d \tag{39}$$

Now it is necessary to calculate the values from irradiation (energy) to irradiance (power), which are obtained by the following expression [20]:

$$G_{D,i}(0) = r_d \cdot N_h \cdot H_{D,h}(0) \tag{40}$$

where r_d is 1 for time intervals of one hour, and for any other time interval is:

$$r_d = 24 \frac{180^\circ}{T} \frac{\cos\omega - \cos\omega_s}{\omega_s \cos\omega_s - \mathrm{sen}\,\omega_s} \tag{41}$$

Performing operations, the final equation for the diffuse irradiance in hourly intervals on a horizontal surface is found:

$$G_{D,i}(0) = r_d \cdot N_h \cdot r_{g24} \cdot H_{g,dm}(0) \cdot k_d \tag{42}$$

3.5. Global Irradiance on Sloped Surface

The most obvious way to calculate the global irradiance on a sloped surface, $G_{g,i}(\beta,\alpha)$, is treat separately each of its components (Figure 2) [3]: direct irradiance, $G_{B,i}(\beta,\alpha)$, the diffuse irradiance, $G_{D,i}(\beta,\alpha)$, and the reflected irradiance, $G_{R,i}(\beta,\alpha)$. Once they are known, the following relationship is applied:

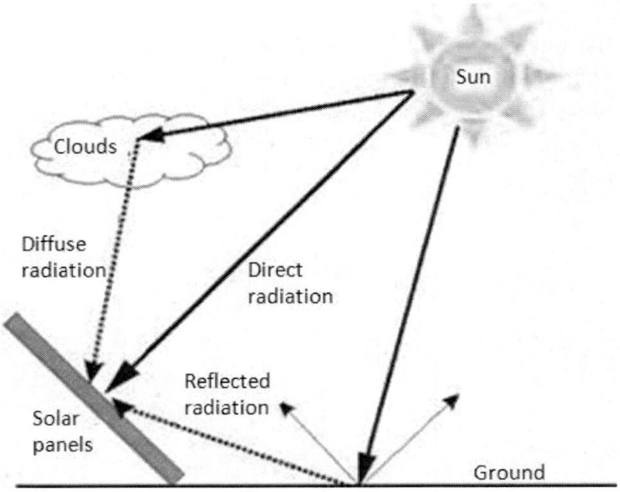

Figure 2. Solar radiation on sloped surface.

$$G_{g,i}(\beta,\alpha)=G_{B,i}(\beta,\alpha)+G_{D,i}(\beta,\alpha)+G_{R,i}(\beta,\alpha) \tag{43}$$

The direct irradiance on a tilted surface can be calculated from the direct irradiance on a horizontal surface through the following expression [3]:

$$G_{B,i}(\beta,\alpha)=G_{B,i}(0)\frac{\max(0,\cos\theta_s)}{\cos\theta_{zs}}FT_B(\theta_s) \tag{44}$$

where $\cos\theta_s$ y $\cos\theta_{zs}$ are parameters depending on geographic location and solar time, whose calculation is exposed in section 2. The factor $FT_B(\theta_s)$ takes into account the possible dirt on solar panels [4] (section 2).

Regarding the diffuse radiation, Hay and Davis [25] proposed to consider the diffuse radiation as a result of an isotropic component and a component circumsolar, weighted according to what they called anisotropy index, k_1:

$$G_{D,i}(\beta,\alpha)=G_{D,i}(0)\left[(1-k_1)\frac{1+\cos\beta}{2}FT_D(\beta)+\frac{k_1\cdot\max(0,\cos\theta_s)}{\cos\theta_{zs}}FT_B(\theta_s)\right] \tag{45}$$

where β is the tilt angle of the solar panels, $FT_D(\beta)$ is another dirt factor [4] whose computation can be seen in section 2. The anisotropy index is defined as:

$$k_1 = \frac{G_B(0)}{B_0(0)} \tag{46}$$

being $G_B(0)$ the deterministic direct irradiance on horizontal surface and $B_0(0)$ the extraterrestrial irradiance on horizontal surface.

With regard to the reflected irradiance, except for snowy environments, the reflectivity on most of grounds is rather low. So that the contribution of albedo to incident radiation on the photovoltaic receptors is usually very small, and it does not justify the development of sophisticated models for its consideration [3]. So it's usually used the simple assumption that the ground is horizontal, infinite and isotropic. Then, the reflected irradiance on a tilted surface is given by:

$$G_{R,i}(\beta,\alpha) = G_{g,i}(0)\cdot\rho\frac{1-\cos\beta}{2}FT_R(\beta) \tag{47}$$

where ρ is the ground reflectivity and $FT_R(\beta)$ is another dirt factor [4] whose determination can be seen in section 2.

In the calculation of the global irradiance on a horizontal surface, $G_{g,i}(0)$, according to equation (43) for a horizontal surface, the reflected horizontal irradiance can be neglected, $G_{R,i}(0)$, since it is practically impossible that the reflected radiation on the ground impinges on the upper side of a horizontal surface, then:

$$G_{g,i}(0) = G_{B,i}(0) + G_{D,i}(0) \rightarrow G_{B,i}(0) = G_{g,i}(0) - G_{D,i}(0) \tag{48}$$

Taking this into account, equation (44) becomes:

$$G_{B,i}(\beta,\alpha) = \left[G_{g,i}(0) - G_{D,i}(0)\right]\frac{\max(0,\cos\theta_s)}{\cos\theta_{zs}}FT_B(\theta_s) \tag{49}$$

Substituting equations (45), (47) and (49) into equation (43), and performing some operations, the expression for the global irradiance on inclined surface is obtained:

$$G_{g,i}(\beta,\alpha) = A\cdot G_{g,i}(0) + B\cdot G_{D,i}(0) \tag{50}$$

in which parameters A and B are:

$$A = \frac{\max(0,\cos\theta_s)}{\cos\theta_{zs}}FT_B(\theta_s) + \rho\frac{1-\cos\beta}{2}FT_R(\beta)$$

$$B = (1-k_1)\frac{1+\cos\beta}{2}FT_D(\beta) + \frac{\max(0,\cos\theta_s)}{\cos\theta_{zs}}FT_B(\theta_s)(k_1-1) \tag{51}$$

If $G_{g,i}(0)$ and $G_{D,i}(0)$ are considered random variables, and their values are substituted into (50) from the expressions (35) and (42) respectively, it's obtained:

$$G_{g,i}(\beta,\alpha) = D' \cdot K_T + F' \cdot k_d \tag{52}$$

where:

$$\begin{aligned} D' &= A \cdot N_h \cdot r_g \cdot B_{0,dm}(0) \\ F' &= B \cdot N_h \cdot r_d \cdot r_{g24} \cdot H_{G,dm}(0) \end{aligned} \tag{53}$$

3.6. Output Electrical Power

When the global irradiance on a tilted surface is known, the calculation of the output electrical real power from the photovoltaic generator is given by the following expression [1,2]:

$$P_{pv} = \eta_c \cdot S \cdot G_{g,i}(\beta,\alpha) \tag{54}$$

where S is the total collection surface of the photovoltaic generator and η_c is the electrical efficiency of the photovoltaic panels. The electrical efficiency of the power inverter isn't considered in this work.

The efficiency of photovoltaic panels is variable and dependent, among other factors, on the global radiation [26], $G_{g,i}(\beta,\alpha)$:

$$\eta_c = \eta_{ref} \left\{ 1 - \beta_{ref} \left[T_a - T_{ref} + \frac{G_{g,i}(\beta,\alpha)}{G_{NOCT}} \left(T_{NOCT} - T_a \right) \right] \right\} \tag{55}$$

where T_{ref}, η_{ref}, β_{ref} are reference temperature, performance and temperature coefficient (°C^{-1}) of the photovoltaic panels, respectively. T_{NOCT} and G_{NOCT} are temperature and irradiance in normal operation conditions of the panels. All these parameters are provided by the manufacturer of the photovoltaic modules. The variable T_a represents to ambient temperature.

Replacing in equation (54) the value of efficiency according to expression (55) is:

$$P_{pv} = N \cdot G_{g,i}(\beta,\alpha) - Q \cdot G_{g,i}^2(\beta,\alpha) \tag{56}$$

where

$$\begin{aligned} N &= S \cdot \eta_{ref} - S \cdot \eta_{ref} \beta_{ref} T_a + S \cdot \eta_{ref} \beta_{ref} T_{ref} \\ Q &= \frac{S \cdot \eta_{ref} \beta_{ref} \left(T_{NOCT} - T_a \right)}{G_{NOCT}} \end{aligned} \tag{57}$$

3.7. Lineal Approximation

It can be seen that (56) is a non-lineal equation. To solve this equation with the method proposed, it is necessary to linearize this equation. The linearization process is made around the solution obtained by a deterministic way, specifically using as input data the expected values of the random variables.

To illustrate the concept of this technique, two random variables *X* and *Y* will be consider, which at some point the problem will be multiplied to give a third random variable *Z*:

$$Z = XY \tag{58}$$

If the deviations of *X* and *Y* around their average values are represented by ΔX and ΔY, respectively, and assuming:

$$X \approx \overline{X} + \Delta X \text{ and } Y \approx \overline{Y} + \Delta Y \tag{59}$$

it is obtained, after neglecting terms of second order $\Delta X \cdot \Delta Y$:

$$Z \approx \overline{X}\overline{Y} + \overline{X}\Delta Y + \overline{Y}\Delta X = \overline{X}Y + \overline{Y}X - \overline{X}\overline{Y} \tag{60}$$

Therefore, if variations of the random variables are small, the variable *Z* can be linearized once the expected values *X* and *Y* are known.

This technique is applied to global irradiance on sloped surface in equation (56). Thus the following expression is obtained:

$$P_{pv} = \left[N - 2Q \cdot \overline{G}_{g,i}(\beta,\alpha)\right] G_{g,i}(\beta,\alpha) + \overline{G}^2_{g,i}(\beta,\alpha) \tag{61}$$

where $\overline{G}_{g,i}(\beta,\alpha)$ is the global irradiance on sloped surface deterministically calculated by equation (52), taking into account the average values of K_T and k_d. These average values are calculated as the first moment around zero, by equations (64) and (65) applied to the probability density functions of K_T and k_d, which are expressed in (18) and (21), respectively.

If it's taken into account the value of $G_{g,i}(\beta,\alpha)$ in equation (52), it can be written the equation for the output electrical power from the photovoltaic generator as follows:

$$P_{pv} = M + D \cdot K_T + F \cdot k_d \tag{62}$$

being:

$$\begin{aligned} M &= \overline{G}^2_{g,i}(\beta,\alpha) \\ D &= \left[N - 2Q \cdot \overline{G}_{g,i}(\beta,\alpha)\right] D' \\ F &= \left[N - 2Q \cdot \overline{G}_{g,i}(\beta,\alpha)\right] F' \end{aligned} \tag{63}$$

By applying to equation (62) the method of cumulants, it is possible to obtain the characteristics of the random variable P_{pv}.

4. Cumulants Method

The cumulants method, which is developed in this section, can replace the convolution of random variables by the sum of their cumulants. Thus, the computational cost is reduced [16,27]. This method also allows using any type of random variable, not just normal distributions.

4.1. Moments and Cumulants

4.1.1. Moments

The moments of a random variable X are the expected values of certain functions of X [28,29]. These are a collection of descriptive measures that can be used to characterize the probability distribution of X and to specify this distribution whether all moments of X are known. Although the moments of X can be defined around any reference point, generally they are defined around zero or the expected value of X. The use of the moments of a random variable to characterize the probability distribution is a very useful task.

Definition 1. The rth moment around zero of X is defined as:

$$\mu_r' = E\left(X^r\right) = \sum_x x^r p(x) \tag{64}$$

if X is discrete and takes values x_i with probabilities $p(x_i)$, or,

$$\mu_r' = E\left(X^r\right) = \int_{-\infty}^{\infty} x^r f(x) dx \tag{65}$$

if X is continuous with density function $f(x)$.

Definition 2. The rth central moment of X or the rth moment around the mean of X is defined as:

$$\mu_r = E(X-\mu)^r = \sum_x (x-\mu)^r p(x) \tag{66}$$

if X is discrete and takes values x_i with probabilities $p(x_i)$, or,

$$\mu_r = E(X-\mu)^r = \int_{-\infty}^{\infty} (x-\mu)^r f(x) dx \tag{67}$$

if X is continuous with density function $f(x)$. μ is the mean of X, i.e., the first around zero moment of the random variable X.

4.1.2. Cumulants

The cumulants of a random variable, such as the moments, are a set of constants that allow to know its properties and to determine its distribution function [30]. However, the cumulants have a number of properties that make them more useful.

The cumulants of a random variable X can be calculated from the central moments of X and vice versa according to the relations which are included in section 4.2. The relations among the moments around zero and the cumulants can also be found in that section.

4.1.2.1. Cumulants Properties

Cumulants are used due to the following advantages, which simplify the calculation of the probability density functions (PDF) and cumulative distribution functions (CDF), for random variables to calculate.

Property I. Let $\xi_1, \xi_2, \ldots, \xi_n$ be independent random variables and $k_\nu^{\xi_1}$, $k_\nu^{\xi_2}, \ldots, k_\nu^{\xi_n}$ their cumulants of order υ, respectively, then, the cumulants of the sum $\varsigma = \xi_1 + \xi_2 + \ldots + \xi_n$ are equal to the sum of the cumulants of each variable,

$$k_\nu^{\varsigma} = k_\nu^{\xi_1} + k_\nu^{\xi_2} + \ldots + k_\nu^{\xi_n} \tag{68}$$

Property II. Let ς be a lineal function of ξ, $\varsigma = a\xi + b$, and k_ν^{ξ} the cumulants of ξ, then, the cumulants of ς, k_ν^{ς}, are:

$$\begin{array}{ll} k_\nu^{\varsigma} = a k_\nu^{\xi} + b & si\,\nu = 1 \\ k_\nu^{\varsigma} = a^\nu k_\nu^{\xi} & ,\ \mathrm{si}\,\nu > 1 \end{array} \tag{69}$$

Property III. Let $\xi_1, \xi_2, \ldots, \xi_n$ be independent random variables and $k_\nu^{\xi_1}$, $k_\nu^{\xi_2}, \ldots, k_\nu^{\xi_n}$ their cumulants of order υ, respectively, then, the cumulants of the sum $\varsigma = a_1\xi_1 + a_2\xi_2 + \ldots + a_n\xi_n + b$ are:

$$\begin{array}{ll} k_\nu^{\varsigma} = a_1 k_\nu^{\xi_1} + a_2 k_\nu^{\xi_2} + \ldots + a_n k_\nu^{\xi_n} + b & si\,\nu = 1 \\ k_\nu^{\varsigma} = a_1^\nu k_\nu^{\xi_1} + a_2^\nu k_\nu^{\xi_2} + \ldots + a_n^\nu k_\nu^{\xi_n} & ,\ \mathrm{si}\,\nu > 1 \end{array} \tag{70}$$

4.2. Relationships among Moments and Cumulants

Various relationships among statistical moments and cumulants of a random variable are shown in this section, which are useful in the application of the cumulants method.

Relationship among the Moments around Zero and the Central Moments

The moments around zero μ'_r of a distribution function can be obtained from the central moments μ_r of this distribution according to the following relationship [30]:

$$\mu'_r = \sum_{k=0}^{r} \frac{r!}{(r-k)!} \mu_k \mu^{r-k} \tag{71}$$

where μ is the mean or expected value of the distribution and r the order of the moment to be calculated.

Calculation of the Cumulants from the Central Moments

The following relationships allow to calculate the first seven cumulants k_1, k_2,...,k_7 of a random variable from its first seven central moments μ_1, μ_2,..., μ_7 [30]:

$$\begin{aligned}
k_1 &= \mu \\
k_2 &= \mu_2 \\
k_3 &= \mu_3 \\
k_4 &= \mu_4 - 3\mu_2^2 \\
k_5 &= \mu_5 - 10\mu_3\mu_2 \\
k_6 &= \mu_6 - 15\mu_4\mu_2 - 10\mu_3^2 + 30\mu_2^3 \\
k_7 &= \mu_7 - 21\mu_5\mu_2 - 35\mu_4\mu_3 + 210\mu_3\mu_2^2
\end{aligned} \tag{72}$$

Calculation of the Central Moments from the Cumulants

The following relationships allow to calculate the first seven central moments μ_1, μ_2,..., μ_7 of a random variable from its first seven cumulants k_1, k_2,..., k_7 [30]:

$$\begin{aligned}
\mu_1 &= 0 \\
\mu_2 &= k_2 \\
\mu_3 &= k_3 \\
\mu_4 &= k_4 + 3k_2^2 \\
\mu_5 &= k_5 + 10k_3k_2 \\
\mu_6 &= k_6 + 15k_4k_2 + 10k_3^2 + 15k_2^3 \\
\mu_7 &= k_7 + 21k_5k_2 + 35k_4k_3 + 105k_3k_2^2
\end{aligned} \tag{73}$$

Calculation of the Cumulants from the Moments around Zero

The following relationships allow to calculate the first seven cumulants k_1, k_2,..., k_7 of a random variable from its first seven around zero moments μ'_1, μ'_2,..., μ'_7 [30]:

$$
\begin{aligned}
k_1 &= \mu_1' \\
k_2 &= \mu_2' - \mu_1'^2 \\
k_3 &= \mu_3' - 3\mu_2'\mu_1' + 2\mu_1'^3 \\
k_4 &= \mu_4' - 4\mu_3'\mu_1' - 3\mu_2'^2 + 12\mu_2'\mu_1'^2 - 6\mu_1'^4 \\
k_5 &= \mu_5' - 5\mu_4'\mu_1' + 10\mu_3'\mu_2' + 20\mu_3'\mu_1'^2 + 30\mu_2'^2\mu_1' + 60\mu_2'\mu_1'^3 - 24\mu_1'^5 \\
k_6 &= \mu_6' - 6\mu_5'\mu_1' + 30\mu_4'\mu_1'^2 - 10\mu_3'^2 + 120\mu_3'\mu_2'\mu_1' - 120\mu_3'\mu_1'^3 + 30\mu_2'^2 - 270\mu_2'^2\mu_1'^2 \\
&\quad + 360\mu_2'\mu_1'^4 - 120\mu_1'^6 \\
k_7 &= \mu_7' - 7\mu_6'\mu_1' - 21\mu_5'\mu_2' + 42\mu_5'\mu_1'^2 - 35\mu_4'\mu_3' + 210\mu_4'\mu_2'\mu_1' - 210\mu_4'\mu_1'^3 \\
&\quad + 140\mu_3'^2\mu_1' + 210\mu_3'\mu_2'^2 - 1260\mu_3'\mu_2'\mu_1' + 840\mu_3'^3\mu_1'^4 - 630\mu_2'^3\mu_1' \\
&\quad + 2520\mu_2'^2\mu_1'^3 - 2520\mu_2'\mu_1'^5 + 720\mu_1'^7
\end{aligned}
\tag{74}
$$

Calculation of the Moments around Zero from the Cumulants

The following relationships allow to calculate the first seven around zero moments μ_1', μ_2',..., μ_7' of a random variable from its first seven cumulants k_1, k_2,..., k_7 [30]:

$$
\begin{aligned}
\mu_1' &= k_1 \\
\mu_2' &= k_2 + k_1^2 \\
\mu_3' &= k_3 + 3k_2k_1 + k_1^3 \\
\mu_4' &= k_4 + 4k_3k_1 + 3k_2^2 + 6k_2k_1^2 + k_1^4 \\
\mu_5' &= k_5 + 5k_4k_1 + 10k_3k_2 + 10k_3k_1^2 + 15k_2^2k_1 + 10k_2k_1^3 + k_1^5 \\
\mu_6' &= k_6 + 6k_5k_1 + 15k_4k_2 + 15k_4k_1^2 + 10k_3^2 + 60k_3k_2k_1 + 20k_3k_1^3 + 15k_2^3 + 45k_2^2k_1^2 \\
&\quad + 15k_2k_1^4 + k_1^6 \\
\mu_7' &= k_7 + 7k_6k_1 + 21k_5k_2 + 21k_5k_1^2 + 35k_4k_3 + 105k_4k_2k_1 + 35k_4k_1^3 + 70k_3^2k_1 \\
&\quad + 105k_3k_2^2 + 210k_3k_2k_1^2 + 35k_3k_1^4 + 105k_2^3k_1 + 105k_2^2k_1^3 + 21k_2k_1^5 + k_1^7
\end{aligned}
\tag{75}
$$

4.3. Resolution Method

The cumulants method is used to solve linear equations systems when the variables involved in the problem are random variables. The method consists basically of obtaining solution cumulants solving the equations system of the problem for each cumulant order of the variables. The cumulants obtained provide the necessary information for rebuilding the probability density functions and cumulative distribution functions of output variables. Let $X_1, X_2, \dots, X_N$ be independent random variables and $Y_1, Y_2, \dots, Y_N$ the unknown random variables in the linear equations system in (76)

$$
\begin{pmatrix} Y_1 \\ Y_2 \\ \vdots \\ Y_N \end{pmatrix} = \begin{pmatrix} a_{11} & a_{12} & \cdots & a_{1N} \\ a_{21} & a_{22} & \cdots & a_{2N} \\ \vdots & \vdots & \ddots & \vdots \\ a_{N1} & a_{N2} & \cdots & a_{NN} \end{pmatrix} \times \begin{pmatrix} X_1 \\ X_2 \\ \vdots \\ X_N \end{pmatrix} + \begin{pmatrix} b_1 \\ b_2 \\ \vdots \\ b_N \end{pmatrix}
\tag{76}
$$

First, for the application of the cumulants method, it is necessary to calculate the cumulants of $X_1, X_2,...,X_N$. Normally these cumulants are calculated from the statistical moments of the random variables using the relationships specified in section 4.2. The cumulants of order v are $k_v^{X_1}, k_v^{X_2},..., k_v^{X_N}$. Applying the property III of cumulants, it is possible to calculate the cumulants $k_v^{Y_1}, k_v^{Y_2},..., k_v^{Y_N}$ of $Y_1, Y_2,...,Y_N$, respectively. Therefore, the system (76)is solved for each order of cumulants, i.e., if it's wished to obtain the first n cumulants of each random variable Y, the following systems are resolved:

$$\begin{pmatrix} k_v^{Y_1} \\ k_v^{Y_2} \\ \vdots \\ k_v^{Y_N} \end{pmatrix} = \begin{pmatrix} a_{11} & a_{12} & \cdots & a_{1N} \\ a_{21} & a_{22} & \cdots & a_{2N} \\ \vdots & \vdots & \ddots & \vdots \\ a_{N1} & a_{N2} & \cdots & a_{NN} \end{pmatrix} \times \begin{pmatrix} k_v^{X_1} \\ k_v^{X_2} \\ \vdots \\ k_v^{X_N} \end{pmatrix} + \begin{pmatrix} b_1 \\ b_2 \\ \vdots \\ b_N \end{pmatrix} \tag{77}$$

for v =1, and

$$\begin{pmatrix} k_v^{Y_1} \\ k_v^{Y_2} \\ \vdots \\ k_v^{Y_N} \end{pmatrix} = \begin{pmatrix} a_{11}^v & a_{12}^v & \cdots & a_{1N}^v \\ a_{21}^v & a_{22}^v & \cdots & a_{2N}^v \\ \vdots & \vdots & \ddots & \vdots \\ a_{N1}^v & a_{N2}^v & \cdots & a_{NN}^v \end{pmatrix} \times \begin{pmatrix} k_v^{X_1} \\ k_v^{X_2} \\ \vdots \\ k_v^{X_N} \end{pmatrix} \tag{78}$$

for $v = 2, 3, \ldots, n$. Then, n systems of linear equations must be solved.

When the cumulants of $Y_1, Y_2,...,Y_N$ are calculated, it can be obtained their respective moments by using the relationships from section 4.2. Both cumulants and moments provide data about the calculated random variables, such as the mean or standard deviation.

It is important to note that for the application of the property III, the variables $X_1, X_2,...,$ X_N, whose linear combination gives the variables $Y_1, Y_2,...,Y_N$, have to be compulsorily independent. The way that this always happens is expressing the unknown variables as a function of the problem data.

When input data $X_1, X_2,...,X_N$ aren't independent, a rotational transformation method is used to transform a set of correlated random inputs ($X_1, X_2,...,X_N$)into a set of uncorrelated random inputs ($X'_1, X'_2,...,X'_N$). This is a first necessary step in the cumulants method when there is some correlation among random inputs. The method is based on the eigenvalues and corresponding eigenvectors of the covariance matrix that can be seen in [31].

In the case of photovoltaic generator the equations system to be solved is reduced to a single equation, given by the expression (62).

Regarding the number of cumulants needed for the reconstruction of the solution, it has been seen in the literature that the number varies from 7 [16] to 9 [27] cumulants. Seven cumulants have been considered enough in this work to illustrate the method proposed.

5. Characterization of Random Variables

When the cumulants method is applied, the results obtained are the cumulants of random variables which are problem unknowns. In this work the unknown variable of the problem is the output power from the photovoltaic generator, P_{pv}.

To characterize these variables is useful to obtain, from the cumulants, probability density functions and/or cumulative distribution functions, and to do on them measurements necessary or graphical representations to have a visual idea of the results.

5.1. Gram-CharlierExpansion

One of the methods used to calculate the cumulative distribution functions and probability density functions of a random variable from its central moments is the Gram-Charlier method. This method is based on the Gram-Charlier expansion which is detailed below.

The Gram-Charlier series of type A use the Hermite polynomials for obtaining an expansion of the density function of a random variable. In practice, only the first terms of this expansion are considered. The literature consulted [30] advises to take into account seven terms. The resulting series truncated may be seen as a normal density function multiplied by a polynomial that takes into account the differences of the calculated density function with respect to a normal distribution.

That is, the Gram-Charlier method gives, from the central moments of a distribution, an approximation based on the normal distribution.

Let ς be a random variable with mean μ and standard deviation σ. According to Gram-Charlier expansion, the cumulative distribution function, $F(x)$ and the probability density function, $f(x)$, of the standardized variable $x = \frac{\varsigma - \mu}{\sigma}$ can be expressed as follows:

$$F(x) = \Phi(x) + \frac{c_1}{1!}\Phi'(x) + \frac{c_2}{2!}\Phi''(x) + \frac{c_3}{3!}\Phi^{(3)}(x) + \ldots \tag{79}$$

$$f(x) = \phi(x) + \frac{c_1}{1!}\phi'(x) + \frac{c_2}{2!}\phi''(x) + \frac{c_3}{3!}\phi^{(3)}(x) + \ldots \tag{80}$$

where $\Phi(x)$ y $\phi(x)$ represent the cumulative distribution function (CDF) and the probability density function (PDF), respectively, of the normal distribution with mean μ=0 and standard deviation σ=1, and $\Phi'(x)$, $\phi'(x)$, $\Phi''(x)$, $\phi''(x)$... their successive derivatives.

The coefficients c_k are constants defined as:

$$c_k = (-1)^k \int_{-\infty}^{\infty} H_k(x) f(x) dx, \qquad k = 1, 2, 3, \ldots \tag{81}$$

where $H_k(x)$ is the Hermite polynomial of order k. In practice, the coefficients of the Gram-Charlier expansion can be expressed in function of the central moments of the object random variable. The first seven coefficients are:

$$\begin{aligned}
c_0 &= 1, \\
c_1 &= c_2 = 0, \\
c_3 &= -\frac{\mu_3}{\sigma^3}, \\
c_4 &= \frac{\mu_4}{\sigma^4} - 3, \\
c_5 &= -\frac{\mu_5}{\sigma^5} + 10\frac{\mu_3}{\sigma^3}, \\
c_6 &= \frac{\mu_6}{\sigma^6} - 15\frac{\mu_4}{\sigma^4} + 30, \\
c_7 &= -\frac{\mu_7}{\sigma^7} - 21\frac{\mu_5}{\sigma^5} + 105\frac{\mu_3}{\sigma^3}
\end{aligned} \tag{82}$$

From the cumulative distribution function (CDF) for a random variable, it is possible to obtain the probability of finding a value within specified limits. Consider Figure 3, the probability that the random variable X is less than the upper limit, and higher than its lower limit, is given by [32]:

$$P\left[L^{low} \le X \le L^{up}\right] = F\left(L^{up}\right) - F\left(L^{low}\right) \tag{83}$$

where X is the random variable, and $F(x)$ is the cumulative distribution function.

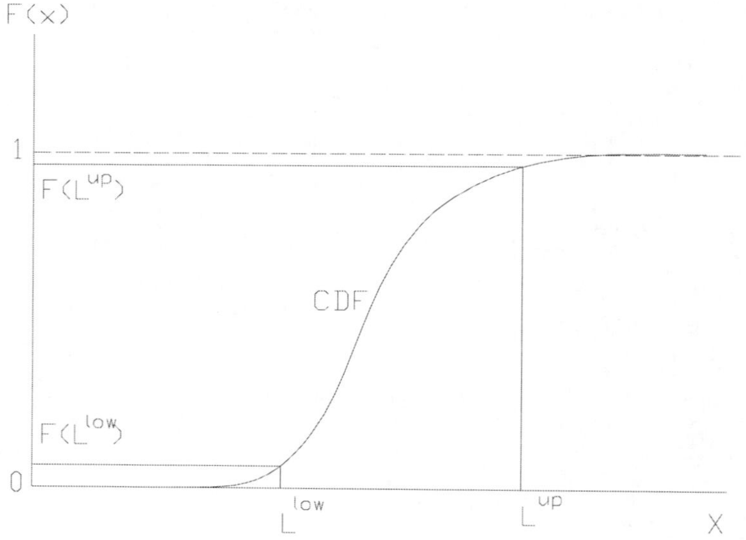

Figure 3. Cumulative distribution function of a random variable X.

In Figure 4 can be seen the process to be followed for determining the output power from a photovoltaic generator as a random variable, by the method above.

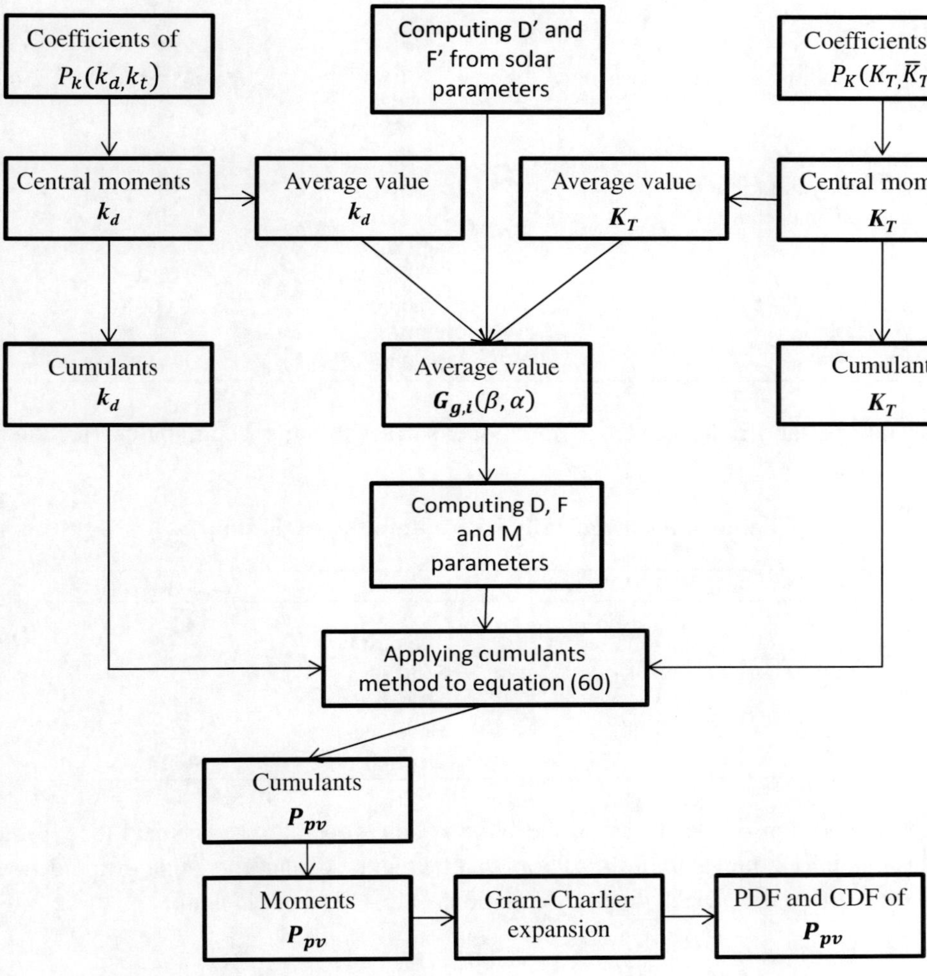

Figure 4. Flowchart of proposed method.

6. Results

The aim is to obtain the probability density function and cumulative distribution function for a photovoltaic generator in two different situations: the first one in a summer day and the second one in a winter day. In the two situations are considered a time interval of 30 minutes.

The photovoltaic generator data and the geographic location in both situations are shown in Table 2.

Table 2. Photovoltaic generator data

	Summer	Winter
Longitude (°)	3	3
Latitude (°)	40	40
Time	11:45-12:15 h	11:45-12:15 h
Month	July	January
Day of the year	185	15
Tilt panels (°)	30	40
Collection surface (m^2)	4500	4500
Ambient temperature (°C)	28	10
T_{ref}	25	25
η_{ref}	0.12	0.12
β_{ref}(°C^{-1})	0.0045	0.0045
T_{NOCT} (°C)	46	46
G_{NOCT}(kW/m^2)	0.8	0.8

Data for global irradiance on a horizontal surface at this geographical location are presented in Table 3.

Table 3. Average daily irradiation for each month

Month	$H_{g,dm}$ (kW/m^2)	Month	$H_{g,dm}$ (kW/m^2)
January	2.6	July	7.6
February	3.3	August	7.0
March	4.7	September	5.7
April	5.4	October	4.0
May	6.7	November	2.9
June	7.4	December	2.1

A good condition of cleanness on the photovoltaic generator is assumed (see Table 1). Under this condition, the value of the factors that take into account the angle of incidence and the degree of dirt on the panels, given by equations (11), (12) and (13), are:

Summer	Winter
$FT_B(\theta_s) = 0.9966$	$FT_B(\theta_s) = 0.9985$
$FT_D(\beta) = 1.0000$	$FT_D(\beta) = 1.0000$
$FT_R(\beta) = 1.0000$	$FT_R(\beta) = 1.0000$

It can be seen as these coefficients are nearly unity. Therefore, given the complexity of their equations, for not very rigorous calculations these coefficients could be neglected, i.e., set them to 1.

In Table 4 first seven central moments and their respective cumulants, for clearness index and the fraction of diffuse radiation are exposed. The moments are determined by applying equation (67) to expressions (18) and (21). The cumulants are calculated from the relations presented in section 4.2.

Table 4.Central moments and cumulants of K_T and k_d

	Winter		Summer	
	Central moments	Cumulants	Central moments	Cumulants
K_T	μ_1=1.1102e-16	k_1=0.5617	μ_1=-1.1102e-16	k_1=0.6728
	μ_2=0.0360	k_2=0.0360	μ_2=0.0177	k_2=0.0177
	μ_3=-0.0052	k_3=-0.0052	μ_3=-0.0029	k_3=-0.0029
	μ_4=0.0038	k_4=-9.7858e-05	μ_4=0.0015	k_4=0.0006
	μ_5=-0.0012	k_5=0.0006	μ_5=-0.0006	k_5=-8.7597e-05
	μ_6=0.0007	k_6=-0.0003	μ_6=0.0003	k_6=-2.7495e-05
	μ_7=-0.0003	k_7=-6.9180e-05	μ_7=-0.0002	k_7=3.9441e-05
k_d	μ_1=0.0755e-15	k_1=0.3499	μ_1=0.0266e-15	k_1=0.1984
	μ_2=0.0233	k_2=0.0233	μ_2=0.0047	k_2=0.0047
	μ_3=0.0072	k_3=0.0072	μ_3=0.0007	k_3=0.0007
	μ_4=0.0027	k_4=0.0011	μ_4=0.0002	k_4=9.8633e-05
	μ_5=0.0011	k_5=-0.0005	μ_5=4.3095e-05	k_5=9.0687e-06
	μ_6=0.0005	k_6=-0.0006	μ_6=1.3273e-05	k_6=-4.7062e-07
	μ_7=0.0002	k_7=-0.0002	μ_7=4.6306e-06	k_7=-4.4249e-07

Before applying the method of cumulants is necessary to know the coefficients for equation (62). These coefficients are expressed in Table 5.

Table 5. Coefficients for equation (62)

Coefficient	Summer	Winter
D	539.0972	725.9065
F	-30.4559	-715.8213
M	0.6255	0.0954

Table 6.Central moments and cumulants of P_{pv}

	Winter		Summer	
	Central moments	Cumulants	Central moments	Cumulants
P_{pv}	μ_1=0	k_1=157,2729	μ_1=0	k_1=353,5494
	μ_2=30924,5088	k_2=30924,5087	μ_2=5044,6938	k_2=5044,6938
	μ_3=-4611720,2745	k_3=-4611720,2745	μ_3=-452817,0232	k_3=-452817,0232
	μ_4=3.1336e9	k_4=2.6462e8	μ_4=1.2516e8	k_4=48813637,9899
	μ_5=-1.1991e12	k_5=2.2705e11	μ_5=-2.6626e10	k_5=-3.7829e9
	μ_6=6.6199e14	k_6=-1.1704e14	μ_6=7.0365e12	k_6=-6.3339e11
	μ_7=-3,4769e17	k_7=1,0652e16	μ_7=-1,8997e15	k_7=4.8465e14

When these coefficients and the cumulants of the clearness index and the fraction of diffuse radiation are known, it is possible to apply the method of cumulants. Then, the cumulants of the output electrical power of the photovoltaic generator are obtained, as shown in Table 6. After that, by applying relations depicted in section 4.2, the central moments of the random variable for the output electric power of the photovoltaic generator are obtained.

Now, Gram-Charlier expansion is applied to obtain the probability density function and cumulative distribution function for P_{pv}. In Figures 5 and 6, these PDFs and CDFs can be seen, respectively.

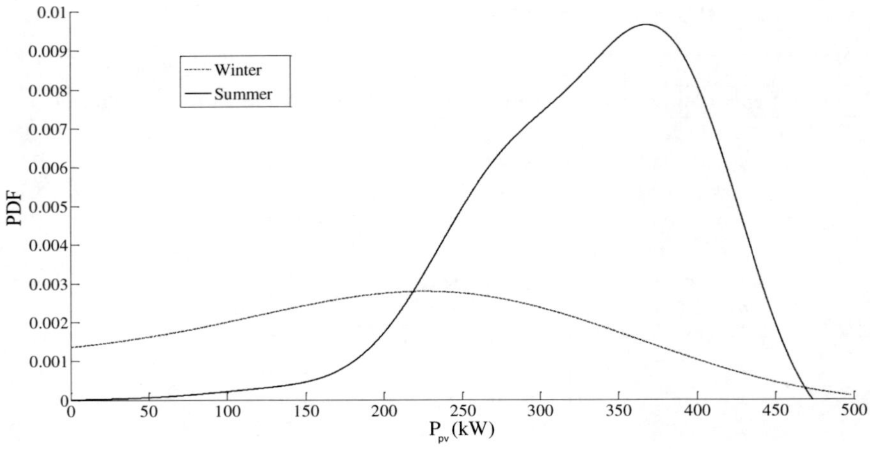

Figure 5. PDFs of P_{pv} in winter and summer.

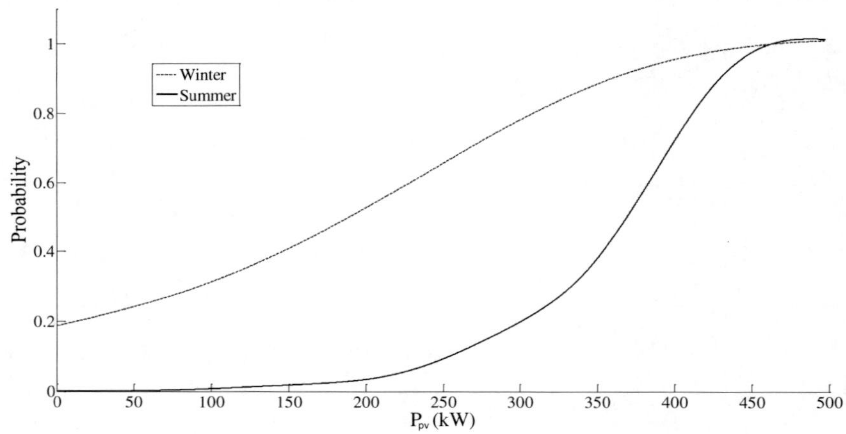

Figure 6. CDFs of P_{pv} in winter and summer.

It is noted that, although mathematically it is possible that these distribution functions have negative values, in practice values higher than zero are only considered, since it is physically meaningless negative values. This means that the photovoltaic generator would performance as a load, which is not considered in this work.

From the CDF of the output electrical power from the photovoltaic generator can be obtained the probability that the generator works in a range of power (equation 83). For example, the probability that the generator operates between limits from 200 kW to 400 kW is:

$$
\begin{aligned}
&F_{P_{pv}}^{Win}\left(400kW\right)=0.9517 \text{ and } F_{P_{pv}}^{Win}\left(200kW\right)=0.5245 \\
&P_{Win}\left(200<P_{pv}<400\right)=F_{P_{pv}}^{Win}\left(400\right)-F_{P_{pv}}^{Win}\left(200\right)=0.4272\rightarrow 42.72\%
\end{aligned}
\tag{84}
$$

$$F_{P_{pv}}^{Sum}(400kW) = 0.7202 \text{ and } F_{P_{pv}}^{Sum}(200kW) = 0.0343$$
$$P_{Sum}(200 < P_{pv} < 400) = F_{P_{pv}}^{Sum}(400) - F_{P_{pv}}^{Sum}(200) = 0.6859 \rightarrow 68.59\% \quad (85)$$

Conclusion

This work has shown how to obtain the output electrical power from a photovoltaic generator as a random variable. The results can be used, for example, to do a study on an electrical system using probabilistic load flow.

This random variable is obtained considering the inherent uncertainties in the climatic conditions, through the clearness index and the fraction of diffuse radiation. Both, the clearness index and the fraction of diffuse radiation, are considered as random variables and their probability density functions are presented.

In order to solve the equation obtained for the output electrical power from of the photovoltaic generator, the method of the cumulants and Gram-Charlier expansion have been used, obtaining satisfactory results, as can be seen in the results section.

The model presented in this work takes into account the random nature of solar radiation, dependent on the geographical location where the solar panels are installed, and the hour and day of the year in which the simulation is performing.

This approach allows determining the output electrical power from a photovoltaic generator at any place and hour as random variable. A study has been performed on a representative summer day and another representative winter day, since photovoltaic generation depends largely on the season of year.

References

[1] Ruiz-Rodriguez F.J., Hernandez J.C., Jurado F. "Probabilistic load flow for radial distribution networks with photovoltaic generators". *IET Renewable Power Generation.* (2012), vol. 6 (2), p. 110 - 121.

[2] Ruiz-Rodriguez F.J., Hernandez J.C., Jurado F. "Probabilistic load flow for photovoltaic distributed generation using the Cornish–Fisher expansion". *Electric Power Systems Research.* 2012, vol. 89, p. 129– 138.

[3] Lorenzo E. Electricidad Solar Fotovoltaica. *Volumen II: Radiación Solar y dispositivos Fotovoltaicos*. Sevilla: Progensa, 2006.

[4] Martín N., Ruiz J.M. "Calculation of the PV modules angular losses under field conditions by means of an analytical model. *Solar Energy Material and Solar Cells.* (2001). vol. 70 (1), p. 25-38.

[5] Hollands K.G.T., Huget R.G. "A probability density function for the clearness index, with applications". *Solar Energy.* (1983), vol. 30 (3), p. 195–209.

[6] Hollands K.G.T., Crha S.J. "A probability density function for the diffuse fraction, with applications". *Solar Energy.* (1987), vol. 38 (4), p. 237–245.

[7] Whillier A. Solar energy collection and its utilization for house heating. *Ph.D. Thesis in Mechanical Engineering,* M.I.T, Cambridge, Massachusetts, 1953.

[8] Liu B.Y.H., Jordan R.C. "The interrelationship and characteristic distribution of direct, diffuse and total solar radiation". *Solar Energy.* (1960), vol. 4 (3), p. 1–19.
[9] Bendt P., Collares-Pereira M., Rabl A. "The frequency distribution of daily isolation values". *Solar Energy.* (1981), vol. 27 (1), p. 1–5.
[10] Saunier G.Y., Reddy T.A., Kumar S. "A monthly probability distribution function of daily global irradiation values appropriate for both tropical and temperate locations". *Solar Energy.* (1987), vol. 38 (3), p. 169–177.
[11] Tiba C., Siqueira A.N., Fraidenraich N. "Cumulative distribution curves of daily clearness index in a southern tropical climate". *Renewable Energy.* (2007), vol. 32 (13), p. 2161–2172.
[12] Orgill F., Hollands K.G.T. "Correlation equation for hourly diffuse radiation on a horizontal surface". *Solar Energy.* (1977), vol. 19 (4), p. 357–359.
[13] Erbs D.G., Klein S.A., Duffie J.A. "Estimation of the diffuse radiation fraction of hourly, daily, and monthly-average global radiation". *Solar Energy.* (1982), vol. 28 (4), p. 293–302.
[14] Reindl D.T., Beckman W.A., Duffie J.A. "Diffuse fraction corrections". *Solar Energy.* (1990), vol. 45 (1), p. 1–7.
[15] Papoulis A., Pillai S. *Probability, Random Variables, and Stochastic Processes,* 4th ed., McGraw-Hill, New York, 2002.
[16] Zhang P., Lee S.T. "Probabilistic load flow computation using the method of combined cumulants and Gram-Charlier expansion", *IEEE Trans. Power Syst.* (2004), vol. 19 (1), p. 676–682.
[17] Kendall M.G., Stuart A. The advanced theory of statistics. *Volume I. Distribution Theory. Charles Griffin and company limited.* London, 1963.
[18] Page J. *Proc. U. N. Conference New Sources Energy* (1961), p. 378-390.
[19] Ruth D.W., Chant R.E."The relationship of diffuse radiation to total radiation in Canada". *Solar Energy.* (1976), vol. 18 (2), p. 153-154.
[20] Collares-Pereira M., Rabl A. "The average distribution of solar radiation--correlations between diffuse and hemispherical and between daily and hourly insolation values". *Solar Energy.* (1979), vol. 22 (2), p. 155-164.
[21] Spencer J.W. "A comparison of methods for estimating hourly diffuse solar radiation from global solar radiation". *Solar Energy.* (1982), vol. 29 (1), p. 19-32.
[22] Rao C.R.N., Bradley W.A., and Lee T.Y. "The diffuse component of the daily global solar irradiation at Corvallis, Oregon (U.S.A.)". *Solar Energy.* (1984), vol. 32 (5), p. 637-642.
[23] Bugler J.M. "The determination of hourly insolation on an inclined plane using a diffuse irradiance model based on hourly measured global horizontal insolation". *Solar Energy.* (1977), vol. 19 (5), p. 477-491.
[24] Iqbal M. "Prediction of hourly diffuse solar radiation from measured hourly global radiation on a horizontal surface". *Solar Energy.* (1980), vol. 24 (5), p. 491-503.
[25] Hay J.E., D. McKay D.C. "Estimating solar irradiance on inclined surfaces: A review and assessment methodologies". *Int. J. Solar Energy.* (1985), vol. 3, p. 203-240.
[26] Skoplaki E., Palyvos J.A. "On the temperature dependence of photovoltaic module electrical performance: A review of efficiency/power correlations". *Solar Energy.* (2009), vol. 83 (5), p. 614–624.

[27] Sanabria I. A., Dillon T. S. "Stochastic power flow using cumulants and Von Mises functions". *International Journal of Electrical Power and Energy Systems.* (1986), vol. 8 (1), p. 47-60.

[28] Canavos G. C. *Applied Probability and Statistical Methods.* Little, Brown and Company, Virginia, 1984.

[29] León-García A. *Probability and random processes for electrical engineering.* Second edition. Addison-Wesley. New York, 1994.

[30] Kendall M.G., Stuart A. *The advanced theory of statistics.* Volume I. Distribution Theory. Charles Griffin and company limited. London, 1963.

[31] Mardia KV, Kent JT, Bibby JM. *"Multivariante analysis".* Academic Press, London, 1997.

[32] Ruiz Rodríguez, F.J. Estudio de la fiabilidad de un sistema de energía eléctrica mediante flujos de carga probabilistas.*Degree project.* Director: Pérez Ruiz J., E.T.S.I.I., University of Málaga (Spain). May, 2005.

In: Photovoltaics
Editor: Monique A. Gill, pp. 121-137
ISBN: 978-1-63117-843-6

Chapter 6

THE EMPIRICAL ANALYSIS OF PROMOTION SCHEMES FOR SOLAR PV: TOWARDS A COMPREHENSIVE FRAMEWORK ON INTERACTIONS BETWEEN ASSESSMENT CRITERIA

***Pablo del Río*[1,*]*, Cristina Peñasco*[1] *and Gustav Resch*[2]
[1]CSIC, Madrid, Spain
[2]Technical University of Vienna, Vienna, Austria

Abstract

Support schemes for solar PV have widely been adopted in the world. There is a voluminous literature on the theoretical and empirical analysis of different instruments to support renewable energy sources in genera and solar PV in particular. This literature has mostly focused on the effectiveness and efficiency criteria, although other criteria have also been included in some analyses. However, what is missing is a comprehensive framework which includes all relevant assessment criteria.

In addition, the assessment criteria have been treated in an isolated manner, although these criteria interact between each other.

Therefore, the aim of this paper is to close those gaps in the literature by providing a comprehensive framework which integrates all the relevant assessment criteria, explicitly considering their interactions. The results are deemed useful in order to undertake future multicriteria analyses of support schemes for solar PV.

1. Introduction

Support schemes for solar PV have widely been adopted in the world. According to the IEA, at least, 82 countries had some type of solar policy by now (IEA 2014). There is a

* E-mail address: pablo.delrio@csic.es, Tel: +0034916022560, Address: Pablo del Río. CSIC, C/Albasanz 26-28, Madrid 28037, Spain. (Corresponding author)

voluminous literature on the theoretical and empirical analysis of different instruments for renewable energy promotion in general and for the support of solar PV electricity in particular. This literature has mostly focused on the effectiveness and efficiency criteria, although other criteria have also been included in some analyses.

However, what is missing is a comprehensive framework which includes all relevant assessment criteria. In addition, the assessment criteria have been treated in an isolated manner, although these criteria interact between each other. Therefore, the aim of this paper is to close those gaps in the literature by providing a comprehensive framework which integrates all the relevant assessment criteria, explicitly considering their interactions. The results are deemed useful in order to undertake future multicriteria analyses of solar PV support schemes.

Different streams of the literature may provide relevant insights on the choice of specific criteria. In order to identify relevant "a priori" criteria and their interactions, we heavily draw on existing concepts from both the environmental economics and the innovation economics literatures, which are deemed relevant in this context[1].

This has been complemented with some insights from other streams of the literature, including the literature on learning effects, the political science literature, the empirical literature on policy support schemes for electricity from renewable energy sources (RES-E) and the literature on European Union (EU) harmonisation of RES-E support schemes.

European Commission documents have also been analysed in order to infer relevant criteria. Furthermore, guidelines in existing policy documents have been considered (Mitchell et al., 2011).

The following section proposes and describes key assessment criteria and justifies their usefulness, providing possible indicators pertaining to those criteria. A discussion on the interactions between different criteria is provided in section 3. Section 4 concludes.

2. Main Criteria for the Assessment of Support Schemes for Solar PV

2.1. Effectiveness

One main criterion on which to judge the success of support schemes for solar PV is obviously the extent to which instruments are effective in triggering deployment. An instrument is said to be effective if it is able to achieve a significant level of solar PV deployment or a certain solar PV target.

In general, effectiveness may refer either to increased generation or increased capacity of solar PV. It can be defined in relative terms: i.e., as a percentage of total electricity or energy consumption (as set in the previous EU Directive 77/2001/EC and in the current Directive 28/2009/EC).

On the other hand, when assessing the effectiveness of a support scheme, the solar PV potentials of countries should be taken into account and the increase in deployment adjusted accordingly. For EU Member States (MS), this is donein the OPTRES, futures-E and RE-

[1] Del Río et al. (2012) provide a detailed discussion of those approaches and identify relevant criteria highlighted by each of them.

Shaping projects, in which the effectiveness of a policy scheme for the promotion of a specific source of renewable electricity such as solar PV is measured as the increase in normalised electricity generation due to this policy compared to the additional available solar PV generation potential or the gross electricity consumption (Ragwitz et al., 2007). More specifically, the effectiveness of a Member State's policy is interpreted as the ratio of the change in the normalised electricity generation over a given period of time and the additional realisable mid-term potential until 2020 for solar PV electricity, where the exact definition of effectiveness reads as follows:

$$E_n = \frac{G_n - G_{n-1}}{ADD - POT_{n-1}}$$

E_nEffectiveness Indicator for Solar PV for the year (n)

G_nElectricity generation potential by Solar PV in year (n)

$ADD - POT_n$ Additional generation potential of Solar PV in year (n) until 2020

This definition of effectiveness has the advantage of giving an unbiased indicator with regard to the available potentials of a specific country for solar PV. Countries need to develop solar PV proportionally to the given potential to show the comparable effectiveness of their instruments (Ragwitz et al., 2007).

However, another, not mutually exclusive definition of effectiveness has proven relevant at least in the context of the EU. This concerns target attain-ment: i.e. the extent to which targets for the penetration of renewable energy in general and solar PV electricity in particular are fulfilled and the trend towards the fulfilment of those targets over time (as in the interim targets in the current EU RES Directive).

2.2. Cost-Effectiveness

Cost-effectiveness generally refers to the achievement of a given solar PV target at the lowest possible cost to society. Environmental Economics sets a clear criterion for cost-effectiveness in reaching a target: i.e. the equimarginality principle. This refers to static efficiency and welfare gains. Cost-effectiveness is attained when an instrument encourages proportionally greater solar PV deployment by those firms and installations with lower deployment costs, and lower deployment by companies with higher deployment costs.

This leads to an equalisation of marginal costs across firms/plants (equimarginality).

The extent to which an instrument encourages the choice of technologies, sizes and places which minimise generation costs is thus a key aspect. This would lead to a minimisation of generation costs across firms/ countries.

Since solar PV has higher generation costs than traditional power generation technologies and also greater than other renewable energy sources, such as wind, they need public support to penetrate the market, which is ultimately paid by consumers and/or taxpayers. While part of the literature has focused on the minimisation of the generation costs and, in fact, some have argued about the need to reduce the overall policy costs for consumers or taxpayers (Huber et al., 2004; Ragwitz et al., 2007; Steinhilber et al., 2011; European Commission, 2008; IEA, 2008a; IEA, 2011). Thus, the costs of support should also be taken into account. Solar PV support is, in the end, generally paid by electricity consumers in their electricity bills.

Therefore, cost-effectiveness has been interpreted in this context as supporting a given amount of solar PV at the lowest possible consumer costs.[2]

In this case, the aim should be to minimise the revenues for producers (to sufficient and appropriate levels)[3].

Figure 1 (below) illustrates the different cost elements.

The transaction costs related to the implementation and functioning of a solar PV support scheme should also be included in the definition of cost-effectiveness. Transaction costs may fall on the public administration or on companies. The former are usually called "administrative costs".

Other costs of solar PV deployment should be taken into account, namely transmission and distribution costs and back-up costs.

System costs include: technology costs (investment costs, capital costs and O and M costs); transmission costs; and back-up costs. System plus policy costs plus transaction (administrative) costs would lead to total costs, as illustrated in Figure 2.

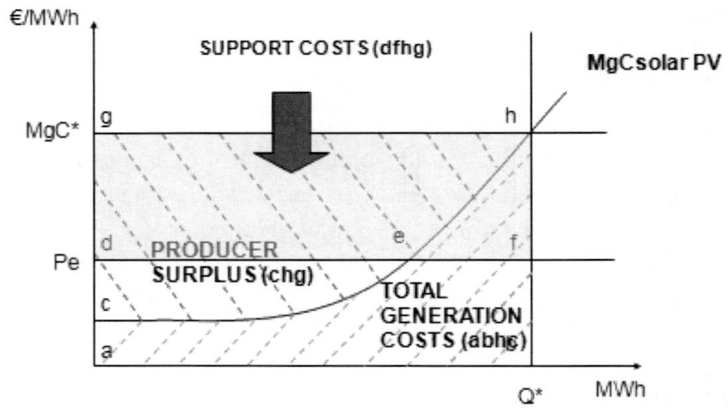

Source: Huber et al. (2004) and Resch et al. (2009).

Note: Q* = Quota or target; MgC* = Marginal costs of the last technology or site needed to comply with the solar PV target/quota. Pe = Wholesale price of electricity. MgCsolar PV = Marginal cost curve of solar PV generation.

Figure 1. Illustrating different cost concepts.

2.3. Dynamic Efficiency

Dynamic efficiency refers to the ability of an instrument to generate a continuous incentive for technical improvements and costs reductions in renewable energy technologies: i.e. an incentive positively to influence technological change processes in the medium and long term. This is a key benefit of investing now in renewable energy technologies because, while renewable energy is not a cost-effective means of reducing CO_2 emissions today, it may be so in the future if investments are made now to accelerate its development. In contrast to the cost-effectiveness criteria, which are much more concerned with the short term, dynamic

[2] See, e.g., Huber et al. (2004), European Commission (2008), Ragwitz et al. (2007), IEA (2008 a), IEA (2011), Mitchell et al. (2011) among others.

[3] Costs for consumers due to solar PV support are defined as transfers from consumers to produ-cers due to solar PV support with respect to the consumer costs due to the purchase of conventional electricity.

efficiency is key in a problem with long-term horizons such as climate change. Future targets regarding GHG emissions and renewable energy are unlikely to be less ambitious than today and, thus, technological change will continue to be a key element in both realms.[4]

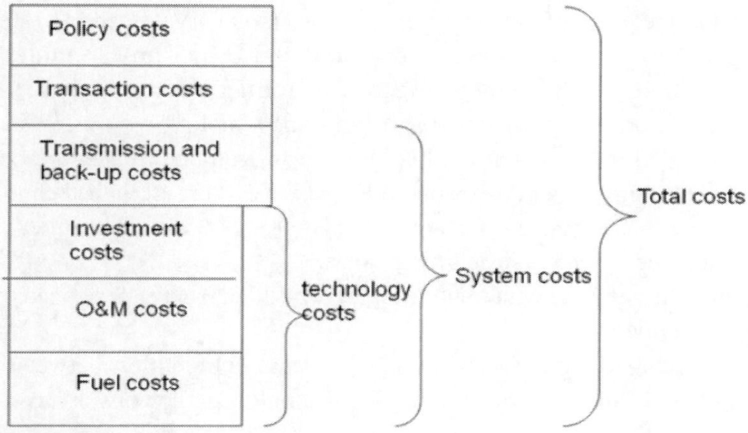

Source: Own elaboration.

Figure 2. Illustrating the different categories of costs.

Those renewable energy support instruments which favour the comercialisation of expensive technologies in niches tend to lead to quality improve-ments and cost reductions; this will allow us to have renewable energy technologies in the future to comply with more ambitious renewable energy and emissions reduction targets at reasonable costs.

If currently expensive mitigation technologies such as solar PV have a large cost reduction potential with increased diffusion (as shown by several studies for energy technologies, see for example (IEA, 2008b), then supporting them today would lead to welfare benefits in terms of intertemporal mitigation efficiency (i.e. cost-effectiveness in the short, medium and long term).

In contrast to cost-effectiveness, dynamic efficiency has an intertemporal perspective on costs.

Solar PV is one of those currently more expensive alternatives.

Instruments for the support of renewable energy which only promote the cheapest and most mature sources would leave solar PV out of the market.

Several authors have emphasised the implications of the path-dependent character of technological change on climate policy (Rip and Kemp, 1998; Unruh, 2000; Marechal, 2007).

If currently expensive technologies with significant potential for quality improvement and costs reduction such as solar PV are not supported today, a vicious circle may ensue: they will remain expensive because they have not been adopted, and they will not be adopted because their high costs make them unattractive for potential adopters.[5]

---

[4] The need for a large-scale deployment of renewables to reduce CO_2 emissions is common in the projections made with simulation models. For example, according to projections made by IEA in its 2008 report on energy technology perspectives (IEA, 2008b), by 2050 the increased use of renewables would contribute 21% to CO_2 emission reductions in the BLUE map scenario (the one compatible with 450ppm concentration levels) with respect to the reference scenario.

[5] The importance of these dynamic efficiency effects is shown by both renewable energy models and climate change models (see, e.g., Stern, 2007).

The impact of renewable energy support schemes upon innovation in renewable energy technologies has several aspects or "dimensions": diversity; R+D; learning effects; and competition (Del Río, 2012). Some are related to other criteria. Diversity refers to the extent to which an instrument favours the deployment of different technologies, including the most mature (wind) and the currently more expensive ones (solar PV). R and D refers to the extent to which a support instrument encourages private R and D by firms. In turn, this is the result of a supply-push and a demand-pull effect. The former occurs because the deployment instrument creates a producer surplus (profit margin) which is reinvested in R and D. The demand-pull relates to the fact that the deployment instrument creates the perspective for a market for the technology, which investors in R and D need to sell their technologies.

Learning effects are related to the effectiveness of the instrument, which allows technologies to advance along their learning curve.

This has proven very relevant with solar PV, which has experienced impressive cost reductions in the last decades.

Finally, the extent to which an instrument favours competition between generators and technology suppliers leading to greater innovation should also be considered.

2.4. Equity

Even if an instrument for the promotion of solar PV leads to net benefits for society as a whole, there will be winners and losers. The distributive impacts upon consumers, citizens, sectors, firms or countries should be considered when designing climate policies at any level (global, European, national or regional). The social acceptance of a given policy depends to some extent upon how those distributive impacts are handled. Distributive concerns are related to winners and losers within countries (consumers, producers and taxpayers) and between countries: i.e. who pays for and who benefits from a given instrument or design element. In particular, it should be identified whether a given instrument leads to a concentration of the costs of solar PV promotion in a limited number of countries.

While minimisation of the total costs of complying with renewable of solar PV targets is part of the cost-effectiveness criterion, compliance costs may fall disproportionally upon countries with lower GDP per capita.

As argued by Capros et al. (2008) for the case of compliance with EU greenhouse gas (GHG) targets, this result was considered by the European Commission to be inconsistent with the equity and fairness criteria which have been set as basic policy principles by the EU.

2.5. Environmental and Economic Effects

The deployment of solar PV projects may bring positive effects for the countries where they are located. Those potential positive effects of solar PV deployment can be included in two categories: environmental and economic effects. The former refers to reduction in GHG emissions and local pollutants, while the latter concerns avoided fossil fuel consumption, which positively affects the trade balance (exports minus imports).

Other co-benefits are also likely, including net job creation, industry creation and exports of renewable energy technology equipment. Finally, it is important to take into account that environmental impacts are not necessarily positive, but may also be negative (visual, land use).

2.6. Socio-Political Feasibility

The implementation of a support scheme which meets all of the aforementioned criteria may still not be socially acceptable and, thus, politically feasible. Social rejection may be of a general nature (i.e., civil society is against the deployment of renewables or against deployment support) or it may have a local character (the so-called 'NIMBY' syndrome).

Compared to other renewable energy technologies (wind and small hydro), the social acceptance of solar PV technologies is greater (European Commission, 2011). They cause a lower NIMBY syndrome.

Likewise, social acceptability is related to the existence of real or perceived local environmental and socioeconomic benefits for specific countries. It may also be related to other criteria. For example, an expensive support scheme is unlikely to be socially acceptable to the general population (consumers). The recent acceleration of solar PV technologies has led to very high increases in the costs of solar PV promotion in some countries (including Spain, Italy, Germany and the Czech Republic), which proved to be socially unacceptable and politically unfeasible, leading to cuts of the support levels for solar PV (sometimes retroactive ones). The (perceived) social acceptability of solar PV policies at the country level can be assumed to translate into a preference of national policy-makers for a specific instrument. Indeed, the political feasibility of a given instrument is related to equity concerns, environmental and economic effects and social acceptability, which may result in significant conflicts with interest groups. Thus, political feasibility deserves separate consideration. Political feasibility depends upon the distribution of the costs of reaching the targets; and awareness of potential local benefits.

3. Interactions between Criteria

While all the criteria considered in this paper are also explicitly or implicitly included in other documents (Mitchell et al., 2011; European Commission, 2008; IEA, 2011), in the literature on renewable electricity support schemes, criteria have traditionally been proposed as a checklist, independent from each other. In reality, however, criteria are interrelated. What is missing is an integrated analysis of how different criteria interact and relate to each other, either directly or indirectly. Only when this is taken into account we can really consider the overall impacts of support schemes on solar PV deployment. This consideration of interactions between criteria is missing in both the solar PV deployment and the renewable electricity deployment literature more generally, probably because most papers follow a disciplinary tradition.

But different perspectives focus on different relevant issues regarding solar PV deployment and they can be and should be combined. In addition, analysing the interactions between criteria is a difficult task. We provide a first attempt identifying possible synergies and conflicts between them. Note that the assessment of solar PV support instruments is affected by the consideration of those interactions between criteria.

A first observation is that the above criteria overlap between each other. This is unavoidable, given their mutual interactions. There is no way in which we can remove one criterion and/or integrate several of them without losing relevant perspectives for the assessment of specific instruments. Criteria are inclusive of all relevant aspects even if this means that one is partially (but never totally) included in others. For example, high consumer costs (cost-effectiveness) affects social acceptability.

But social acceptability also depends on the local benefits of solar PV deployment and on how costs and benefits are distributed among different socioeconomic actors (equity). In turn, the existence of local benefits depends on effectiveness in deployment, which overlaps with dynamic efficiency to create a national industry upstream the innovation process in solar PV technologies.

Finally, political feasibility depends on the one hand on the interaction between social acceptability, cost-effectiveness, local benefits and equity and, on the other hand, on the juridical criteria.

Criteria may certainly be in conflict. For example, a greater level of local benefits at the country or regional level may come at the expense of cost-effectiveness in meeting EU targets. This means that if MS policy makers are interested in the local benefits of renewable electricity, deployment may not occur in those places with a better renewable resource potential in the EU.

Another example of a conflict is between consumer costs and dynamic efficiency. Lower profit margins for solar PV generators would tend to lead to a lower cost for consumers.

But it could also lead to lower incentives for innovation if innovation results from reinvesting the profit that is obtained by solar PV generators into new technologies (developed by equipment producers), although the evidence from the German and Spanish solar PV industry is not so clear in this regard.

In general, a conflict between static and dynamic efficiency could occur if existing, cheaper technologies lock out promising technologies with a large cost reduction potential.

But, on the other hand, there might also be synergies.

For example, effectiveness in deployment of different technologies would encourage dynamic efficiency by facilitating technological diversity and allowing technologies to advance along their learning curves. Furthermore, the existence of a market for the technologies (which is created by renewable electricity and solar PV support instruments) feeds-back into the R and D stage and, thus, deployment triggers R and D investments.

Another example of a synergy between criteria is between static efficiency and political feasibility, in so far as low consumer costs enhance social acceptability and, thus, political feasibility.

In contrast, windfall profits undermine cost-effectiveness, equity, social acceptability and political feasibility. Equity and political feasibility are also obviously interrelated. Note that we have separated social acceptability and political feasibility to grasp relevant interactions between them and other criteria. However, it is very difficult to disentangle both subcriteria. A socially unacceptable instrument is also politically unfeasible.

It may come as a surprise that static efficiency (consumer costs) and effectiveness are positively related through lower investment risks (Mitchell et al., 2006). This is so if a solar PV support scheme which is effective in deployment (because it provides a stable flow of revenues) is also regarded as less risky.

In turn, lower risks obviously entail a lower risk premium and, thus, lower levels of support would be required, which involves lower consumer costs.

Therefore, a holistic perspective on criteria whereby their mutual relations (synergies and conflicts) are made explicit is required. This may help to build a hierarchy of criteria, whereby criteria and subcriteria are related and some are shown to be instrumental in achieving others.

The aim is to come out with a figure identifying those interactions, whereby every link has to be justified according to the aforementioned four types of literatures and contributions. Figure 3 and table 1 picture and summa-rise those interactions. Further details are provided below.

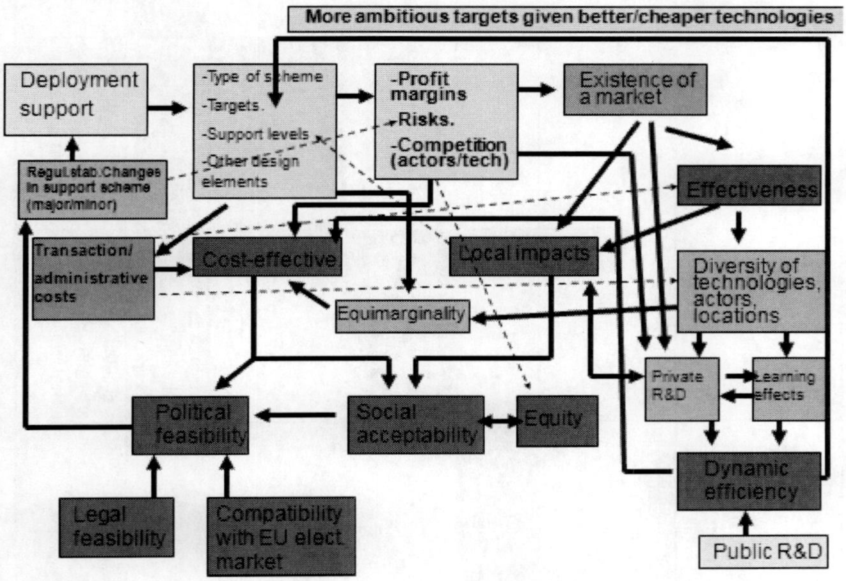

Source: Own elaboration.

Figure 3. Picturing the interactions between criteria.

The upper part of the figure identifies key aspects of support schemes which subsequently trigger a set of relevant effects (profit margins, risks, competition between technologies and between actors, transaction and administrative costs or the equimarginality rule). Several key aspects and features of support schemes are relevant in this context. The type of support scheme is obviously one. The specific design elements of the instrument are another. For example, in the past, some renewable electricity support instruments (feed-in tariffs, FITs) have been shown to be less risky than others (tradable green certificates, TGCs) (Butler and Neuhoff, 2008; Lipp, 2007; Mitchell et al., 2006; European Commission, 2008; Burer and Wustenhagen, 2009; Steinhilber et al., 2011; Ernst and Young, 2010). TGCs have also shown to lead to greater profit margins for more mature technologies (European Commission, 2008) and for incumbent actors (Verbruggen, 2009; Bergek and Jacobsson, 2010), while less mature/more expensive technologies are not allowed to penetrate the market. In contrast, profit margins tend to be greater for immature technologies under FITs compared to TGCs, since the TGC price has been lower than the marginal costs of many immature/expensive technologies and, particularly, solar PV (Ragwitz et al., 2007).

Table 1. Illustrating the interactions between criteria

From (columns)/to (rows)	Effectiveness	Cost efficiency	Dynamic efficiency	Equity	Local impacts	Social acceptance	Political feasibity
Effectiveness						(indirect effect through political feasibility)	Regulatory stability as a result of political feasibility favors deployment
Cost efficiency			Innovation positively influences cost-effectiveness In t1 (techno cost reductions)				Regulatory stability results in lower risk premium
Dynamic efficiency	Market creation leading to learning effects and private R and D						
Equity					Local impacts have equity effects some of which are difficult to predict		
Local impacts	Deployment leads to local impacts		Creation of a local industry and impacts upstream the innovation process (techno. diversity).				Indirectly through impact of political feasibility on effect.
Social acceptance		Greater consumer costs reduce social accept.		Distributive impacts of the support scheme affects social accept.	Benefits of solar PV deployment results in social accept.		
Political feasibility		High consumer costs make continuation of support scheme unlikely		Unequitable schemes are politically unfeasible in the long-term	Greater local benefits make the continuation of support politically feasible	Social accept. is a crucial element of political feasibility	

Source: Own elaboration based on Del Río et al. (2012).

In turn, profit margins are positively related to R and D investments (see below).

In addition, TGCs have traditionally been assumed to promote greater levels of technological competition (European Commission, 1999) although, in reality, all instruments promote competition whether by reducing costs (TGCs) or by increasing revenues (FITs). Furthermore, competition is enhanced when there are few entry barriers into the market and many and diverse actors enter the market.

FITs have also proven superior to TGCs in this regard, since TGCs have mostly favoured the large generators, possibly due to the higher transaction costs of the instrument falling disproportionately on smaller generators.

The type of instrument and its design elements obviously affect the transaction costs of the instruments and their administrative costs, although not much empirical research has been undertaken in the renewable electricity support literature on this topic. Both transaction and administrative costs affect the total costs of an instrument (i.e., static efficiency) but transaction costs also have an impact on effectiveness and diversity of actors and technologies, since they fall disproportionately on the smaller actors (given their large fixed-cost component). The type of support scheme and design element, targets and support levels affect profit margins, risk and the level of competition. This has a bearing on effectiveness in deployment and on the cost-effectiveness of the scheme, in this later case through their impact on policy costs. However, they may also have a direct effect on cost-effectiveness through their impact on the equimarginality rule which leads to a minimisation of generation costs.

Effectiveness (market creation) is a crucial criterion which is clearly instrumental in achieving others, since it triggers multiple effects.

Two are worth mentioning: local benefits and dynamic efficiency.

Regarding the former, it is well known that renewable energy deployment in general and solar PV deployment in particular has significant local/national impacts.

Some may be negative (i.e., negative externalities in the form of impacts or soil occupancy or negative impact on grid stability), while others are positive (including job creation, rural and regional development opportunities and diversification of energy supply).

These local benefits are crucial for the acceptability of the solar PV support scheme by the general population of a country and, thus, for its politi-cal feasibility (see below).

In turn, one local benefit is the creation of a local industry. This leads to the existence of domestic suppliers of the technology which, in turn, affect policies and, particularly, support levels (lobbies).

Building a constituency behind the new technology (advocacy coalitions) has positive effects on several criteria (social acceptability, R and D in dynamic efficiency) but may also lead to regulatory capture, rent seeking and, thus, negatively affect cost-effectiveness.

Concerning dynamic efficiency, there are feedback loops from diffusion to the previous stages of the innovation process. Private R and D can be influenced by support instruments in the form of deployment incentives although it is certainly not the only source of R and D, the other being public investments in R and D in the form of direct subsidies, tax incentives, tax credits etc.

Indeed, the innovation literature has often stressed the complementary role of private and public R and D in the innovation process, although their relative importance may vary along the different stages[6]. The link between deployment support and private R and D investments takes place through two mechanisms: profit margins and the existence of a market. The profit margins which are particularly relevant for R and D investments are those obtained by immature, higher-cost technologies, with a greater improvement potential through R and D, whereas excessive profit margins to mature technologies (such as wind energy) are more likely to result in windfall profits[7]. Private R and D also benefits from the existence of a local manufacturing industry which, in turn, highly depends on the effectiveness of the support scheme.

A second major source of dynamic efficiency is the activation of the different types of learning effects, which generally takes place as a result of effectiveness in deployment. Note that private R and D and learning effects interact (Watanabe et al., 2000). A lower cost for technologies as a result of R and D makes them more attractive for potential adopters, increases their diffusion and allows them to advance faster along their learning curve. On the other hand, learning reduces costs and promotes diffusion. In turn, market creation makes RD and D investments in those technologies more attractive.

Therefore, it becomes obvious from this analysis that effectiveness (market creation) is instrumental for the dynamic efficiency criterion (both regarding R and D and learning effects). The recently developed functional approach with the systems of innovation perspective provides the theoretical and empirical underpinning for this link, by stressing the crucial role of the "market creation" function in the different renewable technology innovation systems (TIS), including the solar PV TIS.

In a long-term perspective, dynamic efficiency relates to other relevant criteria and affects crucial aspects of the support scheme. For example, better and/or cheaper technologies as a result of R and D investments would allow the setting of more stringent renewable electricity targets in the future or the reduction of support levels over time.

Furthermore, dynamic efficiency impacts are crucial for the cost-effectiveness of the scheme in a long-term perspective.

On the other hand, the role of competition should be acknowledged in this context. Competition is given much emphasis in the traditional economics literature and it is usually considered to be a source of innovation.

Since competition between actors and technologies is usually stressed, the usual recommendation is for renewable electricity support instruments which favour "technological neutrality" and avoid "winner picking" such as in unbanded TGC schemes. There are some problems with this view, however.

First, competition should not be an end in itself, but is rather instrumental in achieving the other criteria. Competition exists as long as a market has been created for market participants to compete, i.e., when a market for the techno-logy has been created. Overemphasis on technology neutrality (which is assumed to lead to competition) rules out in effect the existence of a market for immature and/or expensive renewable energy technologies (such as solar PV) and, thus, competition within this market.

In addition, competition has a qualitative and a quantitative aspect in terms of a diversity of actors. It is favoured when there are many actors in the supply and demand sides. Indeed, competition refers to the number of players in the market (Butler and Neuhoff, 2008; Sawin,

[6] Public support for R and D is particularly necessary where a market failure in the innovation process is more likely to occur, i.e., in basic R and D, whereas it becomes less relevant as we move towards later stages.

[7] Windfall profits may occur both in mature and immature technologies. However, high profit margins in immature technologies are more likely to be reinvested in R and D than is the case with mature technologies.

2004). Paradoxically, instruments overstressing competition and technology neutrality may lead to too few competition levels since the number of actors is effectively restricted. This has been the case with TGC schemes, which have mostly favoured large utilities and incumbents in Flanders (Verbruggen, 2009), Sweden (Jacobsson et al., 2009; Bergek and Jacobsson, 2010) and the UK (Mitchell et al., 2006; Verbruggen, 2009). In contrast, in most countries FITs have shown to favour a diversity of actors (small and large firms) and new entrants into the market.

In turn, FITs have facilitated the creation of supply and demand sides for the technologies in the market, simultaneously providing supply-push and demand-pull influences. FITs have encouraged development of domestic manufacturing industries, which has led to a large number of companies and thereby has created competition (Sawin, 2004)[8].

This is in contrast to technology-neutral schemes which provide a (weak) demand-pull influence with an underdevelopment of the supply side and have hardly generated industry creation.

Finally, some authors have argued that competition occurs independently of the support instrument chosen (Haas et al., 2011). Competitive pressures leading to innovation occur regardless of the support instrument chosen.

Under technology-neutral schemes, the incentive is to manufacture cheaper technologies (so that they can receive TGCs), such as wind electricity.

In contrast, under FITs the incentive is to produce better technologies with increased production efficiencies in order to maximise revenue flows. There is no wonder that the best renewable electricity (and solar PV) equipment manufacturers are traditionally in countries with FITs, whereas technology providers from countries with TGC schemes are scarce.

Cost minimisation is mostly the goal for renewable electricity generators under TGC schemes, whereas the aim is to maximise revenues under FITs.

On the other hand, social acceptability of solar PV deployment and public support for this deployment is a crucial criterion, mostly stressed by the systems of innovation perspective and the political science literature, and confirmed by several empirical studies.

It is closely linked to political feasibility, since in democratic systems policy makers seeking re-election should avoid major conflicts with social constituencies related to solar PV deployment and support. The continuation over time of the support scheme and the support levels themselves depend on social acceptability and political feasibility.

Social acceptability is directly affected by other criteria, i.e., the magnitude of consumer costs and how the costs and benefits are distributed among the population and between countries.

High (or, more importantly, highly increasing) consumer costs as a result of a solar PV support policy are likely to trigger a backlash against the instrument and maybe against solar PV deployment itself. For example, this has recently been shown in the case of solar PV promotion in Spain and also in other European countries.

---

[8] Recently, Mitchell et al. (2011) have also related greater and broader participation to competition: "Inclusive, meaning that potential for participation is as broad as possible on both 1) the supply side (traditional producers, distributors of technologies or energy supplies, whether electricity, heat or fuel), and 2) the demand side (businesses, households etc.), which can'self-generate' with distributed renewable electricity, enabling broader participation that unleashes more capital for investment, helps to build broader public support for renewable electricity (as in Denmark and Germany) and creates greater competition" (op.cit., p.71).

On the other hand, social acceptability is enhanced by the existence of local benefits stemming from solar PV deployment and, particularly, the more visible ones: industry creation and jobs. In reality, creation of a market (effectiveness), local impacts (leading to the creation of a local industry) and social acceptability interact and are likely to generate a reinforcing effect with positive feedbacks, mostly due to advocacy coalitions.

Of course, there might also be negative local impacts in the form of negative environmental externalities due to the concentration of renewable energy projects in general and solar PV projects in particular. This may negatively affect the social legitimacy of renewable electricity and solar PV support (see, for example, Bergmann et al. (2006)) for the case of wind electricity in the UK). Deployment (effectiveness) and social acceptability (NIMBY) may be negatively related at high solar PV penetration levels. Thus, social acceptability becomes proportionally more important with increasing solar PV penetration due to NIMBY and greater costs. Social acceptability influences political feasibility, which is also directly affected by the cost-effectiveness of support, with high-costs of support being politically non-acceptable. Finally, it is worth mentioning that the circular flows in figure 3 provide an inherent dynamic perspective on the interactions between criteria with changes over time. In this context, political feasibility feeds back to solar PV support.

Governments may need to fine tune the solar PV support scheme (either targets, instruments or design elements) as a result of identified drawbacks in the scheme or due to pressures from socioeconomic actors (lobbies, advocacy coalitions). This change in the support scheme would have influences on the criteria in successive time periods. Thus, figure 3 provides a picture of circular flows, and suggests an inherent dynamic perspective on the interactions between criteria with changes over time.

Conclusion

This paper has provided a holistic perspective on criteria and subcriteria for the assessment of solar PV support schemes, mostly with an European Union focus.

In fact, this framework is currently being applied in the analysis of policy pathways for a harmonisation of renewable electricity support schemes in the EU after 2020[9]. Although in this paper we have focused on solar PV in the EU, the framework developed here can be applied to any renewable electricity technology and any country in the world.

In fact, it can be very useful for multicriteria analyses on the topic in so far as the interactions between criteria are made explicit. In the past, assessment criteria were considered isolated from each other, which is not realistic.

From a policy perspective, considering the interaction between criteria is useful in so far as a given policy may simultaneously influence several criteria, something that public policy-makers should take into account when proposing policy measures.

Obviously, our framework has not attempted to identify which criterion is more important than others. The importance attached to different criteria is likely to be different by different stakeholders and, even within the policy-makers group, there are likely to be many differences depending on the administrative level considered (supranational, national, regional or local).

[9] See the Beyond 2020 project website for further details: www.res-policy-beyond2020.eu.

Only relevant stakeholders in a given setting can judge this relevance. This issue should be undertaken in future research. Future research could also aim at applying the developed framework to non-EU contexts and, particular-ly, developing countries, trying to assess its validity in those contexts.

Acknowledgments

The work in this chapter is based on work carried out in the Beyond 2020 project (Design and impact of a harmonised policy for renewable electricity in Europe). This initiative could be established thanks to the financial and intellectual support offered by the "Intelligent Energy – Europe" (IEE) programme of the European Commission, operated by the Executive Agency for Competitiveness and Innovation. Cristina Peñasco gratefully acknowledges support from the Spanish Ministry of Education and CSIC through the JAE program.

References

Bergek, A., Jacobsson, S., 2010. Are Tradable Green Certificates a cost-efficient policy driving technical change or a rent-generating machine? Lessons from Sweden 2003–2008. *Energy Policy*, 38(3), 1255-1271.

Bergmann, A., Hanley, N., Wright, R. 2006. Valuing the attributes of renewable energy investments. *Energy Policy*, 34(9), 1004-1014.

Burer, M. J., Wustenhagen, R. 2009. Which renewable energy policy is a venture capitalist's best friend? Empirical evidence from a survey of international cleantechinvestors. *Energy Policy* 37, 4997–5006.

Butler, L., Neuhoff, K., 2008. Comparison of Feed-in Tariff, Quota and Auction Mechanisms to Support Wind Power Development. *Renewable Energy,* 33, 1854-1867.

Capros, P., Mantzos, L., Papandreu, V., Tasios, N., 2008. Model-based Analysis of the 2008 EU Policy Package on Climate Change and Renewables. *Report to the European Commission – DG ENV*. June 2008.

Del Río, P., 2012. The dynamic efficiency of feed-in tariffs: The impact of different design elements. *Energy Policy*, 41, 139-151.

Del Río, P., Ragwitz, M., Steinhilber, S., Resch, G., Busch, S., Klessmann, C., De Lovinfosse, I., Nysten, J. V., Fouquet, D., 2012. Assessment criteria for identifying the main alternatives. *D2.2 Report under the Beyond 2020 project*, funded by the Intelligent Energy – Europe program http://www. res-policy-beyond2020.eu/

Ernst and Young, 2010. Ernst and Young Country Attractiveness Indices and Renewable Energy Strategy. *Wind Energy Convention 2010*. Madrid, 15 June.

European Commission (EC), 1999. Electricity from renewable energy sources and the internal electricity market. *Commission Working Document*. 13. 04.1999. SEC (1999) 470 final.

European Commission (EC), 2008. SEC(2008) *57 Commission Staff Working Document - The support of electricity from renewable energy sources*. [Online] Available at:http://ec.europa.eu/ [Accessed 07 August 2013].

European Commission (EC), 2011. Public Awareness and Acceptance of CO2 capture and storage. *Special Eurobarometer 364. Wave 75.1*. European Commission.

Haas, R., Resch, G., Panzer, C., Busch, S., Ragwitz, M., Held, A., 2011. Efficiency and effectiveness of promotion systems for electricity generation from renewable energy sources – Lessons from EU countries. *Energy*, 4, 2186-2193.

Huber, C., Faber, T., Haas, R., Resch, G., Green, J., Ölz, S., White, S., Cleijne, H., Ruijgrok, W., Morthorst, P., Skytte, K., Gual, M., del Río, P., Hernández, F., Tacsir, A., Ragwitz, M., Schleich, J., Orasch, W., Boke-mann, M., Lins, C., 2004. Action Plan for deriving dynamic RES-E policies. *Report of the project Green-X*. http://www.greenx.at/downloads/Action%20plan%20for%20deriving%20dynamic%20RESE%20policies%20-%20Green-X.pdf.

IEA (International Energy Agency), 2008a. *Deploying renewables*. Paris.

IEA (International Energy Agency), 2008b. *Energy Technology Perspectives*. Paris.

IEA (International Energy Agency), 2011. *Deploying renewables*. Paris.

IEA (International Enegy Agency), 2014. *IEA/IRENA Global Renewable Energy Policies and Measures Database*. http://www.iea.org/policiesand measures/renewableenergy/Last Access February 2014.

Jacobsson, S., Bergek, A., Finon, D., Lauber, V., Mitchell, C., Toke, D., Verbruggen, A., 2009. EU renewable energy support policy: Faith or facts?. *Energy Policy*, 6, 2143-2146.

Lipp, J., 2007. Lessons for effective renewable electricity policy from Denmark, Germany and the United Kingdom. *Energy Policy*, 35, 5481–5495.

Marechal, K., 2007. The Economics of climate change and the change for climate in economics. *Energy Policy*, 35(10), 5181-5194.

Mitchell, C., Bauknecht, D., Connor, P., 2006. Effectiveness through risk reduction: a comparison of the renewable obligation in England and Wales and the feed-in system in Germany. *Energy Policy*, 34, 297–305.

Mitchell, C., Sawin, J. L., Pokharel, G. R., Kammen, D., Wang, Z., Fifita, S., Jaccard, M., Langniss, O., Lucas, H., Nadai, A., Trujillo Blanco, R., Usher, E., Verbruggen, A., Wustenhagen, R., and Yamaguchi, K., 2011. Policy, Financing and Implementation. In: *IPCC Special Report on Renewable Energy Sources and Climate Change Mitigation* [O. Edenhofer, R. Pichs-Madruga, Y. Sokona, K. Seyboth, P. Matschoss, S. Kadner, T. Zwickel, P. Eickemeier, G. Hansen, S. Schlomer, C. von Stechow (eds.)], Cambridge, Cambridge University Press, available at http://srren.ipcc-wg3.de/report/IPCC_SRREN_Ch11.pdf.

Ragwitz, M., Held, A. Resch, G., Faber, T., Haas, R., Huber, C., Coenraads, R., Voogt, M., Reece, G., Morthorst, P. E., Jensen, S. G., Konstantina-viciute, I., Heyder, B., 2007. *OPTRES – Assessment and optimisation of renewable energysupport schemes in the European electricity market*. Supported by the European Commission (D.G. Energy and Transport). Brussels.

Resch, G., Ragwitz, M., Faber, T., Panzer, C., Haas, R., 2009. 20% RES by 2020. *An assessment of the new EU RES policy framework 10th IAEE European Conference*. Vienna.

Rip, A., Kemp, R., 1998. Technological Change. In: *Human Choice and Climate Change* [S. Rayner, E. Malone (Eds.)], Battelle Press, Columbus, Ohio, Volume 2, 327-399.

Sawin, J., 2004. *Mainstreaming Renewable Energy in the 21st Century*. World Watch Institute, Washington, DC.

Steinhilber, S., Ragwitz, M., Rathmann, M., Klessmann, C., Noothout, P., 2011. D17 Report: Indicators assessing the performance of renewable energy support policies in 27 Member States. RE-Shaping: Shaping an effective and efficient European renewable energy market.

Stern, N., 2007. *The Economics of Climate Change: The Stern Review*. Cambridge University Press.

Unruh, G. C., 2000. Understanding carbon lock-in. *Energy Policy*, 28(12), 817-830.

Verbruggen, A., 2009. Performance evaluation of renewable energy support policies, applied on Flanders' tradable certificates system. *Energy Policy*, 37 (4), 1385–1394.

Watanabe, C., Wakabayashi, K., Miyazawa, T., 2000. Industrial dynamism and the creation of a virtuous cycle between R and D, market growth and price reduction. The case of Photovoltaic Power Generation (PV) development in Japan. *Technovation*, 20(6), 299-312.

In: Photovoltaics
Editor: Monique A. Gill, pp. 139-155
ISBN: 978-1-63117-843-6

Chapter 7

INDOOR PHOTOVOLTAICS: BASIC PRINCIPLES, CURRENT STATUS AND FUTURE DEVELOPMENT

Monika Freunek (Müller), Ph. D.*
TCF Engineering, Germany

Abstract

The use of photovoltaic devices within buildings is also known as Indoor Photovoltaics (IPV). IPV has gained rising interest in the last years, when both the development of energy efficient electronics with an average power demand of a few micro watt and the introduction of narrow banded artificial light sources took place. This has opened new application fields for IPV, such as to power wireless sensor systems in intelligent buildings, with limiting efficiencies close to 50% for fluorescent lighting and even higher for LED spectra. This chapter provides the basic theory, recent progress, and future aspects of IPV devices. Aspects of IPV characterization are explained and referenced to methods from Standard Test Conditions following IEC 60904-3 Ed. 2 (2008). Typical indoor irradiance conditions and the resulting theoretical photovoltaic efficiency limits for ideal and real cells under these conditions are discussed. The calculated limits are compared to recent progress in research and industrial devices. The chapter closes with a discussion of current and future optimization strategies.

PACS: 88.40.hj, 88.40.mr

Keywords: indoor photovoltaics; photovoltaic efficiency; indoor irradiance; low intensity; micro energy harvesting

1. Introduction

Photovoltaics for use in indoor environments is also known as indoor photovoltaics (IPV). IPV devices usually measure a few centimeters and need to operate at intensities between 0.1..10 $\mathrm{Wm^{-2}}$ [1],[2] obtained from solar radiation filtered through a window, diffuse radiation reflected from the surrounding indoor environment, and artificial light sources, such as incandescent bulbs and nowadays fluorescent tubes or light emitting diodes (LED).

*E-mail address: monika.freunek@tcf-engineering.com

The first generation of IPV products has been introduced in the late Seventies. These devices typically consisted of amorphous silicon photovoltaics and are still well known from calculators [3]. Other examples include postage weighing scales [4], kitchen scales, and office table decorations. All of these products were designed for use with incandescent bulbs and required an active orientation to a radiation source from their users. As it has been shown also for use of light switches in schools and office buildings [5], [1], users tend to a passive behavior, which could be described as leave and forget behavior. The acceptance of IPV products therefore remained low with IPV being a niche market with very few research.

The technical possibilities and feasible applications for IPV have changed significantly in the last decade when two major developments took place. First, thermal radiators, such as incandescent bulbs and halogen lamps, which emit below 5% of visible radiation, were widely replaced by efficient narrow banded lighting technologies, such as fluorescent and energy saving lamps and LED. In some regions of the world as for example the Euopean Union with its Regulation 32009R0244, the use of incandescent lamps has even been banned officially [6]. Second, the average electric power need of small electronic devices was reduced to the microwatt scale with demonstrated standby power in the picowatt range [7]. This enabled the concept of micro energy harvesting, i.e. to power electronic devices, such as wireless sensor nodes, using ambient energy, such as kinetic, chemical, thermal or electromagnetic energy [8]. In buildings, optical energy is for the majority of applications the most important ambient energy due to high power densities and good predictability. The major part of applications of IPV will be found in building automation including ambient assisted living, green buildings, or intelligent buildings. These applications require a large set of environmental and local information in order to enable functions such as room temperature control, presence detection or medical informations. In these cases, IPV will be combined with storage devices such as capacitors, ultracaps or accumulators in order to enable maintenance free operation for a typical operation time of ten years.

Consumer electronic devices, such as mobile phones, iPods or tablet PC require significantly more power than micro energy harvesting systems. Their operational power ranges from 10 mW for the standby of a mobile phone to 10..100 W for a tablet PC [9]. However, these products also have a good user acceptance for active recharging strategies, and optimized IPV devices can be used in order to extend the operation time until a charging process with connection to the electrical grid is necessary.

Fig.1 shows exemplary current IPV products. As almost all other commercial products available, they are based on amorphous silicon. Other materials currently used in industry are crystalline silicon and dye sensitized photovoltaic converters. Up to now, materials investigated in research for use in IPV include CdTe, CIGS, CIS, Cu(In,Ga)Se2, GaAs, GaAsP, GaInP, GaAs/GaInP multijunction cells, and P3HT:PCBM.

The efficiency of photovoltaic devices depends on the spectal distribution and radiation intensity of the application environment. In contrast to most other fields of photovoltaics there is no characterization standard for IPV devices yet, even not concerning the reference spectrum. Many methods used in industry and research lack reproducibility, which increases the risk of design errors and product failure. Thus, the most reliable method at the point of writing this chapter is to use a retraceable reference cell, an indoor light source of known spectral distribution and to calibrate this setup by calculating the spectral mismatch between the indoor light source and the reference spectrum of the cell [1].

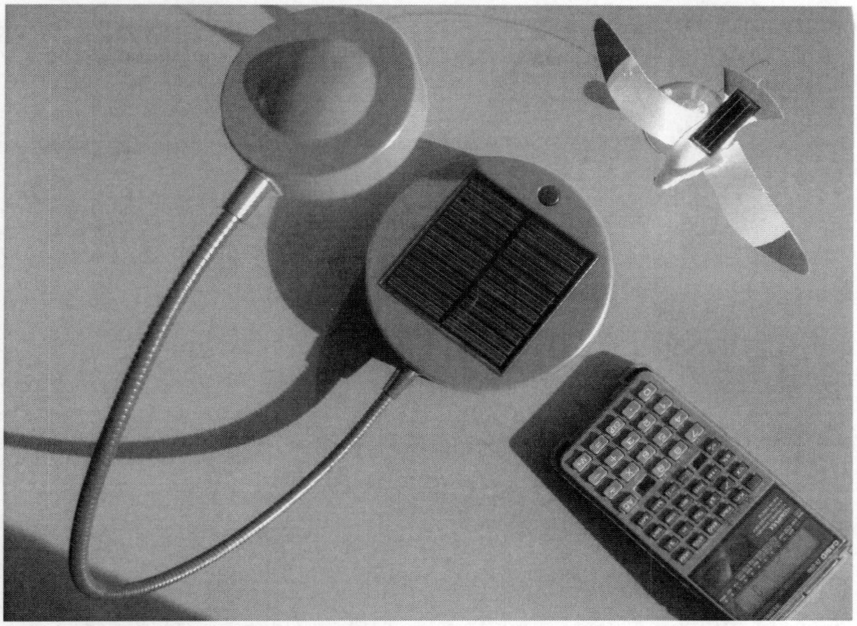

Figure 1. Exemplary indoor photovoltaic products: An office lamp with LED light, a calculator, and a bird for decoration purpose, that moves its wings when light falls on its back.

This chapter is structured as follows. The first section explains the basic theory of indoor photovoltaic devices including their maximum efficiencies for different indoor lighting sources and scenarios, effects of intensity, basic design aspects, and the systematics of available characterization methods. An overview on the current status of indoor photovoltaics in industry and research is given in the second section. Finally, an outlook on future developments of IPV is provided in the third section.

2. Basic Principles

2.1. Maximum Efficiencies

The efficiency η of a photovoltaic converter is given by the ratio of its electrical output power at the operation point (maximum power point) P_{mpp} to the received optical power P_{opt} as given in Eq. 1.

$$\eta = \frac{P_{mpp}}{P_{opt}}, \tag{1}$$

with P_{opt} resulting from

$$P_{opt} = hc \int_0^\infty \frac{\Phi_{p,\lambda}}{\lambda} d\lambda \tag{2}$$

The spectral photon flux per area A of a radiation source depends on the number of incident photons N per second and wavelength. The photon energy E is calculated for

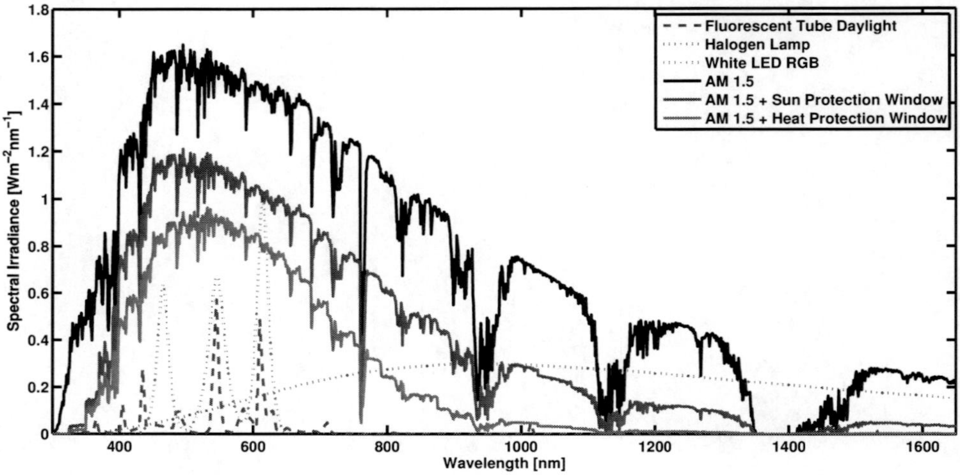

Figure 2. Exemplary spectral distributions of natural and artificial radiation sources.

each wavelength following $E = \frac{hc}{\lambda}$, with $c = 2.998 * 10^8$ ms^{-1} as the speed of light, and $h = 6.626 * 10^{-34}$Js denoting Planck's constant.

Typically, once an application has been chosen, practical efficiency calculations mainly focus on variations of the intensity level, i.e. the optical power received per area. However, for the basic considerations of possible efficiencies, the spectrum of the application will decide on the choice of converter material. Exemplary spectral distributions of natural and artificial radiation sources are depicted in Fig. 2. The well known curve of the solar radiation is the calculated spectrum for passing a reference atmosphere with an air mass factor (AM) 1.5, corresponding to an angle of incidence for the solar radiation of 48°, that is received on a horizontal area with a tilt of 37°to the solar radiation. This spectrum has a power density of 1003 Wm^{-2}. It is defined in IEC 60904-3 Ed. 2 (2008) and the most important reference spectrum for photovoltaic applications [10]. Possible effects of a transmission of solar radiation through a window are shown in the grey lines. Each window has a spectral transmittance $T(\lambda)$. Depending on the latitude and on the purpose of the building, $T(\lambda)$ will be optimized for sun or heat protection as well as for application specific requirements, such as noise or radio wave protection.

The medium grey dotted line represents the spectrum of a thermal radiator for artificial lighting, a halogen lamp. As the sensitivity function of the human eye ranges only from 380..780 nm, the inefficiency of thermal radiators for lighting purpose is obvious.

The other grey lines show exemplary spectra of modern narrow banded lighting technologies, i.e. fluorerescent tubes and light emitting diodes (LED). The spectrum of energy saving lamps is similar to the one of fluorerescent tubes. It clearly follows from Eq. 1 and Eq. 2, that the obtainable electrical power and resulting photovoltaic efficiency obtainable for all these spectra will differ from the results obtained for the AM1.5 solar spectrum.

A common aproach to determine the limiting efficiency of a photovoltaic cell has been presented by Shockley and Queisser [11]. They considered a photovoltaic cell in thermal equilibrium with its environment, that has infinite dimensions, is normally orientated to the radiation, and absorbs every photon of energy E greater than or equal to its energy gap E_g,

$E \geq E_g$. This is equal to an external quantum efficiency EQE = 1. For real cells, the EQE of the converter depends on the choice of materials, and the overall optical and electrical cell design. The resulting electrical current density induced by a spectral photon flux $\Phi_{p,\lambda}$ is the photocurrent density j_{ph} with

$$j_{ph} = q \int_0^\infty \Phi_{p,\lambda} EQE(\lambda) d\lambda, \tag{3}$$

where $q = 1.602 * 10^{-19}$ As is the elementary charge. They also assumed the short circuit current density j_{sc} to be approximately the same as the photocurrent density, $j_{sc} \approx j_{ph}$. Taking into account the typical indoor irradiance conditions, i.e. the spectrum consisting of artificial light sources and solar radiation filtered by a window of a spectral transmission $T(\lambda)$, Eq. 3 needs to be modified as in the following equation [12].

$$j_{ph} = q \int_0^\infty d\lambda EQE(\lambda)[\Phi_{p,\lambda}^{solar} T(\lambda) + \Phi_{p,\lambda}^{artificial}] \tag{4}$$

In most application cases of indoor photovoltaics, the light source that is normal to the photovoltaic converter can be considered as the major light source and should serve as reference spectrum to all design questions. That said, unlike in Shockley Queissers assumptions, there is usually a set of radiation sources to be considered and in many application cases, indirect radiation reflected by the walls, furniture and so on can be the major source of radiation. This is especially valid for low irradiance applications with a typical irradiance ranging between 0.1..1 Wm^{-2}. In order to calculate the indirect radiation, a spectral reflection modell depending on the materials needs to be chosen as for example described in [13]. The simplified approach from Eq. 5 summarizes the solid angles dependencies by introducing a spectral total reflection factor $R(\lambda)$ [12].

$$j_{ph} = q \int_0^\infty \int_0^\omega d\lambda d\Omega EQE(\lambda, \Omega) R(\lambda, \Omega)[\Phi_{p,\lambda}^{solar} T(\lambda, \Omega)) + \Phi_{p,\lambda,\Omega}^{artificiall}] \tag{5}$$

Indoor irradiance values are up to four orders of magnitude below outside conditions. The typical range is between 0.1..10 Wm^{-2}, with maximum values in the 100 Wm^{-2} range for installation directly behind windows [1]. The maximum power density p_{mpp} of a photvoltaic converter is calculated from

$$p_{mpp} = j_{sc} V_{oc} FF \tag{6}$$

The fill factor FF rates the ideality of the cell. The open circuit voltage V_{oc} depends on the cell temperature T, and the ratio of the short circuit current density j_{sc} to the saturation current density j_o:

$$V_{oc} = \frac{k_B T}{q} \ln(\frac{j_{sc}}{j_0} + 1). \tag{7}$$

The first ratio with Boltzmann's constant $k_B = 8.617 * 10^{-5} eVK^{-1}$ is also known as thermal voltage. For the dark current I_o, it follows from the assumptions made by Shockley and Queisser, that losses are only caused by the thermal emission of the cell, a modeling

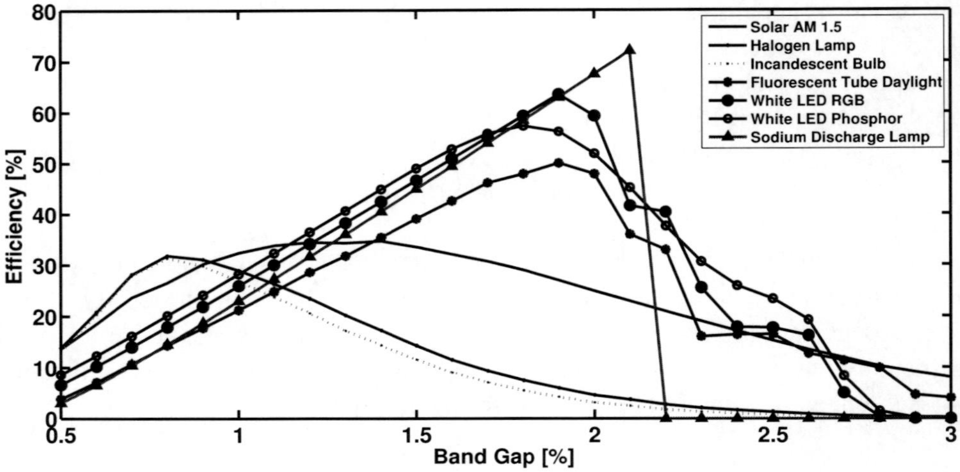

Figure 3. Energy gap E_g and limiting Shockley-Queisser efficiency for the spectral photon flow of a fluorescent tube with daylight spectrum (OSRAM Lumilux 840/18W), terrestrial solar light with standard spectrum AM1.5, a halogen lamp, an incandescent bulb, a RGB white LED, a phosphor white LED, and a sodium discharge lamp [2], [15].

approach which was known as model of detailed balance at that time. This can be approximated following [14].

$$I_0 = \frac{2\pi k_B T q}{h^3 c^2}(E_g + k_B T)^2 \exp(\frac{-E_g}{k_B T}). \tag{8}$$

It can be seen from Eq. 8, that besides the dependence on the operation temperature, it is the band gap, that influences these losses. In principle, increasing the band gap increases both fill factor and open circuit voltage. Fig. 3 shows the maximum photovoltaic efficiencies calculated as outlined above limit for different indoor lighting sources, and, as a reference, for the terrestrial standard spectrum ASTM-G173 AM1.5 global [2], [15].

Planckian or thermal radiators, such as the sun or incandescent and halogen lamps, emit over a broad sprectral range. Thus, for the terrestrial solar light (AM1.5) less than 30% of the incident photonflux contributes to the photovoltaic short circuit current, which gives a limiting efficiency around 33%. The effect of filtering the solar radiation through a window is given in Fig.4 with resulting limiting efficiencies of 40% for an exemplary heat protection window, and of 43% for a sun proctection window. This increase of efficiency is due to the optimization of the spectral transmission of the window to the sensitivity function of the human eye, which leads to a narrowing of the incoming spectrum. As the maximum of the human visibility function is located at a wavenlength of 555 nm, the ideal band gaps for indoor applications based on solar radiation shift to higher values, i.e. 1.4 eV, and 1.7 eV in the given examples.

For halogen lamps and incandescent bulbs the photovoltaic efficiency limit decreases below the limit for terrestrial solar radiation, and the optimum band gap shifts to 0.8 eV, which is close to the Germanium band gap of 0.67 eV.

The spectrum of modern artificial indoor lighting sources is optimized for the visible

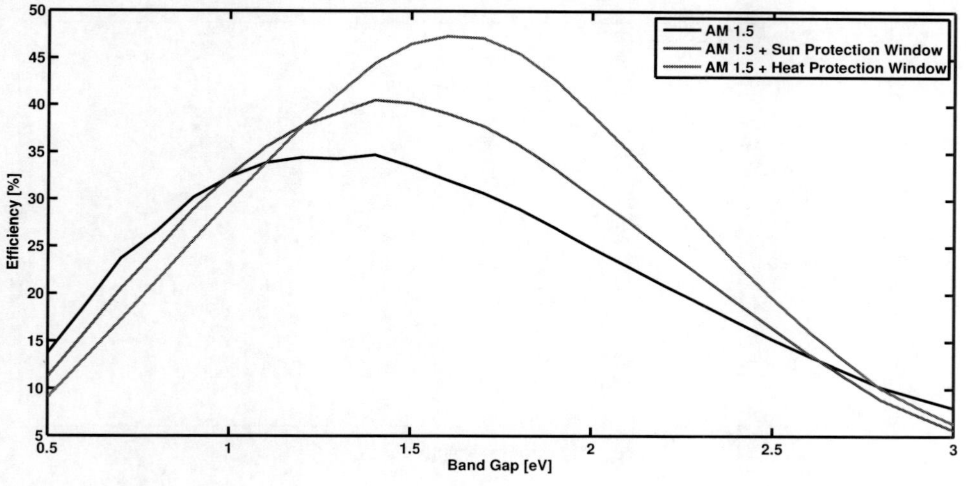

Figure 4. Energy gap E_{g} and limiting Shockley-Queisser efficiency for the spectral photon flow of terrestrial solar light with standard spectrum AM1.5 and the same spectrum transmitted through different types of windows.

range from 380 nm to 780 nm. Therefore it is possible in IPV to shift to larger band gaps with the benefit of having a higher open circuit voltage V_{oc}, while simultaneously having a high photon yield due to the narrow banded spectrum. This results in ideal photovoltaic band gaps around 1.9..2.0 eV and maximum efficiencies of 50% for a fluorescent tube with daylight spectrum with a photon yield close to 90%, 64% efficiency for a RGB white LED, and even 72% for a sodium discharge lamp with monochromatic emission at λ= 589 nm.

The use of laser radiation in industrial settings is an example of IPV applications where a matching of artificial light source and photovoltaic converter is possible at the system design stage. The ideal case of this would be the matching of the energy gap E_{g} of the photovoltaic converter to a monochromatic light source emitting at a wavelength λ, and with an energy $E(\lambda) = E_g$. The efficiency then is [15]

$$\eta = \frac{E_g \int_{E_g}^{\infty} \Phi_{p,\lambda} dE}{\int_0^{\infty} E(\lambda) \Phi_{p,\lambda} dE} \tag{9}$$

The ideal band gap energy of this model is just below the photon energy, as only photons of $E(\lambda) > E_g$ can deliver work. The efficiencies for a monochromatic radiation source of a radiation power P = 1 W calculated following the Shockley-Queisser model are depicted in Fig.5. For E_{g} = 2.3 eV corresponding to an emission wavelength of 555 nm η reaches 73% [15]. The decrease of radiative recombination current in this model leads to the increase of efficiency with the band gap.

2.2. Intensity Effects

Most indoor photovoltaic devices need to operate under low injection conditions. This affects the electrical characteristics, and the appropriate approaches for modeling and design of the devices.

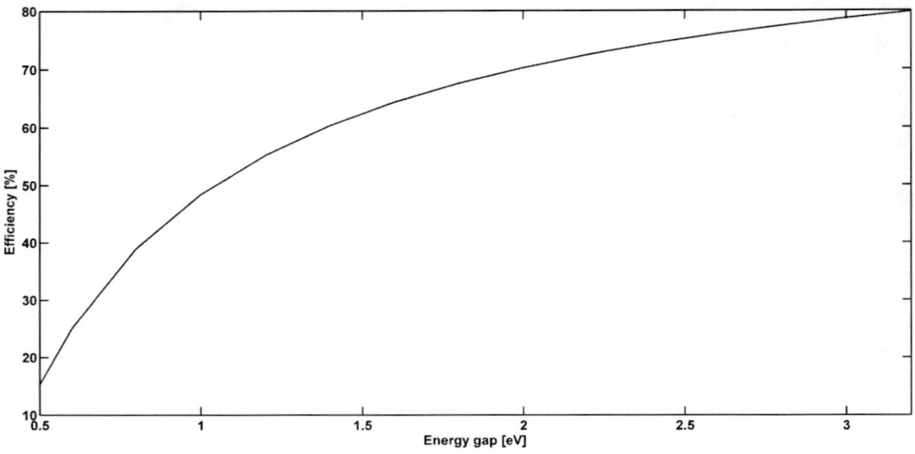

Figure 5. Energy gap E_{g} and limiting Shockley-Queisser efficiency for monochromatic radiation[15].

First, typical indoor intensities and thus the short circuit current are up to four orders of magnitudes below outside conditions. Following Eq.7 this leads also to a slight reduction of V_{oc}. Second, the total dark current I_0 can reach the range of I_{sc} provided a poor design. In this case, the reduction of both open circuit voltage and fill factor can lead to the electrical breakdown of a cell indoors that is fully functional under outside conditions. An example of this can be found in Table 2 in the section below. For indoor and low intensity applications, the reduction of dark current losses is therefore an important design parameter. The total dark current I_0 depends on fundamental properties of the material of choice, such as the band gap and the doping concentrations, on the operation conditions, i.e. temperature and intensity, and on the manufacturing processes during the production of the photovoltaic converter.

The most important losses in semiconductors of indirect band gap as for example silicon are surface recombinations and recombinations in defect states, the so called Shockley-Read-Hall recombinations. They can be minimized by surface passivation and by decreasing the defect state densities. Further losses include drift currents and phonon assisted recombination (Auger recombination). Semiconductors of direct band gap, such as GaInP, are less affected by Shockley-Read-Hall recombination, and do not suffer from Auger losses. Their most important loss mechanism for indoor applications is the emission of photons (radiative recombination current), that is depending on the radiation intensity. Hence, this current is significant for power beam applications using laser radiation.

The total electric losses of a photovoltaic cooverter are often lumped together as a series resistance R_s and a shunt resistance R_{shunt}, which yields [16]

$$\ln\left(\frac{I+I_{ph}}{I_0}-\frac{V-IR_s}{I_0 R_{shunt}}+1\right)=\frac{q}{k_B T}(V-IR_s). \tag{10}$$

Provided a careful design and besides for power beam applications, the effect of the series resistance can be neglected for indoor applications. As low irradiance affects both open circuit voltage and fill factor [17], the shunt resistance needs to be maximized for these applications. As a rule of thumb, the order of magnitude of the required shunt resistance can be approximated by [15]

$$R^{min}_{shunt} \approx \frac{V_{oc}}{I_{sc}}. \tag{11}$$

Thus a shunt resistance of a few Ohm is sufficient for an AM 1.5 global spectrum with an intensity of 1000 Wm^{-2} with an ideal short circuit current of 45 mA for Si (17 mA for GaP). This increases to the range of kΩ for a 4 Wm^{-2} fluorescent spectrum with an ideal short circuit current of 186 μA (165 μA). For some materials, the variance of their shunt resistance with intensity needs to be considered for indoor photovoltaic design.

2.3. Characerization

For the design of reliable IPV products and the comparison of indoor efficiencies in research and industry, standardized measurement conditions are required. Such a standard is available for outdoor solar cells with the IEC 60904-3, Ed. 2 (2008), also known as Standard Test Conditions (STC) [10]. The test conditions include a solar AM 1.5 spectrum, an intensity of 1000 Wm^{-2}, and a cell temperature of 25 °C. For indoor photovoltaic devices, there are no standard test conditions available yet. Instead, there are three main strategies in research and in order to report indoor perfomance values: characterization following STC, photometric indoor measurements, and radiometric characterization under indoor conditions.

Due to the lack of indoor standards, results obtained under STC are also found on datasheets of IPV products. This has the advantage of being reproducible and retraceable to SI units, i.e. to the World Photovoltaic Scale [18]. However, as outlined above, photovoltaic efficiency results obtained under STC are of low predictive value for a perfomance under both low irradiance and artificial light sources. A variation of this is to provide the electrical perfomance values for different intensities. From this, information on the voltage stability can be gained, if the intensity has been lowered to at least 10 Wm^{-2}. Irradiance values greater than this do not match the typical IPV operating conditions, and the effect of the shunt resistance is too low.

A very common method is to measure under an illumination of 200 Lux, i.e. to use a photometric characterization setup. Photometric measurements refer to an illuminance E_v rather than an irradiance E_e. The illuminance E_v for a given irradiance E_e can be calculated by multiplication with the human visibility function $V(\lambda)$ and the luminous coefficient $K_m = 683$lm/W.

$$E_v = K_m \int_{380nm}^{780nm} E_e V(\lambda) d\lambda \tag{12}$$

An illuminance is measured with a luxmeter, which is a semiconductor sensor with a filter mimicking the human visibility function. For historical reasons, luxmeters are calibrated to Class A incandescent bulbs. For use with any other light source, the mismatch factor (MMF) based on the spectral mismatch calculation is to be determined. The spectral mismatch factor MMF for two different spectra $E_{e,ref}$ and E_e and the corresponding short circuit current of the semiconductor device is

$$MMF = \frac{\int_0^\infty I_{sc,ref} E_{e,ref} d\lambda}{\int_0^\infty I_{sc} E_e d\lambda}. \tag{13}$$

The MMF of luxmeters depends on the applied filter. With respect to modern indoor lighting sources, companies often refer to fluorescent tubes, which are the most common office lamps. However, unlike the Class A spectrum, there are many fluorescent spectra. In order to be reproducible, IPV performance data obtained under photometric conditions would need to provide the spectral data or to state the exact type of light source and luxmeter used for the measurements.

A retraceable radiometric characterization setup for IPV with white fluorescent light and measurement intensities of 9.1 Wm^{-2} and 0.1 Wm^{-2} has been demonstrated in research [19], [2]. The setup was calibrated using a high efficiency crystalline silicon solar cell, that had been optimized for low irradiance applications. The short circuit current and its spectral response was measured under STC, and, for the same spectrum and cell temperature, under an intensity of 9.1 Wm^{-2}. The spectral irradiance of the indoor characterization setup light sources, two fluorescent tubes (OSRAM Lumilux 840/18 W daylight), was regulated electrically and determined with a spectrometer for different intensities. The lamps had been chosen due their common use in office buildings and their similar spectrum to energy saving lamps. Based on Eq. 13, the setup was calibrated to 9.1 Wm^{-2}, and 0.1 Wm^{-2}, respectively. For the electrical characterization of IPV devices, a Keithley 6514 Electrometer and a Keithley 2400 SourceMeter were used due to the low currents in the range of microampere. The device temperature was 25 ± 2 °C as described in the IEC 60904-3 Ed.2 (2008).

The measurements of available indoor irradiance follow the same principles. Currently, there are no radiometric measurement devices for IPV available. In principle, the use of semiconductors with a calibrated photocurrent is possible. Thus, silicon irradiance sensors are very common instruments for outdoor applications. However, the sensitivity and the accuracy of the available products is usually too low for indoor environments.

Another common irradiance measurement instrument for outdoor characterization is a pyranometer. Pyranometers consists of thermoelectric arrays, and have a typical spectral response from 280..4000 nm with a hemispheric field of view. The price of pyranometers ranges between a few hundred and a few thousand Dollar, depending on the required accuracy. Pyranometers can be used for indoor irradiance measurements provided the signal to noise ratio is sufficient.

A cost-effective and reliable method is the use of the reference cells following the spectral mismatch procedure described above and provided the cell is calibrated to the application spectrum and its sensitivity is sufficient for low irradiance conditions.

3. Current Status

3.1. Products

Currently, most products employ amorphous silicon with first products using dye sensitized cells. The IPV market is still immature and very dynamic with few companies having a permanent market presence. This can be seen in Table 1, which is an updated version of an IPV product survey published 2009 [21]. Some of the companies have omitted their IPV program due to recent changes in their total photovoltaic business, such as Schott Solar, other companies have been closed, and new companies, especially in the field of dye sensitized cells, have arised.

It can also be seen, that the characterization conditions referred to by the manufacturers are very heterogene, and most of them are not reproducible without further information. IPV designers therefore require own measurements. This is especially valid for the use of dye sensitized devices, where currently no perfomance data are provided from the manufacturers. Manufacturers of c-Si products and flexible devices, which have larger areas and are often designed for outdoor applications, perform their characterization under STC or similar conditions. The manufacturers of the small scale a-Si devices refer to 200 Lux. In principle, only stabilized performance parameters should be stated for a-Si, although under low irradiance conditions the Staebler-Wronski-effect will affect the efficiency less than outdoors. It is important to understand, that the power densities calculated in this table are not comparable because amongst other the information on the spectral distribution of the characterization light source is missing. The results of radiometric and retraceable efficiency measurements of IPV devices are given in the following section.

3.2. Measured Efficiencies

Table 2 provides a listing of efficiencies of photovoltaic cells and modules in the centimeter scale, that were measured under STC and using the radiometric characterization setup with fluorescent tubes with an intensity of 9.1 Wm^{-2} [15], [35]. As besides for a-Si, there are few products available, many of the investigated devices were research prototypes and/or designed for other applications. Amongst others this can be seen for the GaInP device, that had been designed for Concentrated Photovoltaics, or the GaAsP cell, which was a photodiode. However, the relative change of efficiency of all devices follow the predictions obtained from the efficiency discussion outlined above. The performance under indoor conditions of all c-Si devices decreased around 50% compared to the results obtained under STC, with a maximum indoor efficiency of 8.2% for research prototypes from the Fraunhofer Insititute for Solar Energy Systems. The absolute efficiencies of these and of the IXYS sample groups outperformed all a-Si samples by at least 0.5%.

The efficiencies measured for the a-Si devices showed a large spread even for different samples of the same product [2]. This can be an effect of the missing standard for the characterization conditions which also affects the design process. The typical increase of efficiency when operating under indoor conditions was measured to be approximately 1% for the a-Si devices.

The field of dye sensitized and organic photovoltaic converters is of special interest for IPV due to low production cost and high band gaps. The dye sensitized cell more than doubled its performance to 6.8% indoors from 3% under STC. The P3HT:PCBM sample groups showed a slight increase from 3.1% to 3.6% [15] and a competitive efficiency of 5.5% indoors from 2.5% under STC in a recent study [35]. All samples were research prototypes from the Fraunhofer Insititute for Solar Energy Systems and the FMF.

The highest photovoltaic efficiencies measured under fluorescent light are at the time of writing 15.8% for a GaInP cell, followed by a CdTe cell achieving 10.9%, and a GaAsP

Table 1. Commercially available IPV cells and modules

Manufacturer	Product	Technology	Area [cm^{-2}]	Power Density [$mWcm^{-2}$]
Irradiance 1000 Wm^{-2}, terrestrial solar spectrum				
EPS Soltec [22]	SM 02-700	c-Si	108.4 [21]	12.9 [21] [1]
Flexcell*[21]	Sunpack 7W	a-Si, flexible	3150.0	2.1 [2]
HighFlex [23]	HF40	c-Si, flexible	3177.0	12.6 [3]
IXYS [24]	SLMD121H04	c-Si Si	6.0	14.8 [2]
Konarka* [21]	KT 25	organic	201.2	1.1 [3]
Plastecs[25]	a-Si, flexible	SP1.5-50-3	21.1	3.9 [1]
PowerFilm [26]	a-Si, flexible	PowerFilm SP3-37	23.4	2.8 [4]
Sharp[27]	a-Si	LR0GC02	27.7	10.8 [3]
SolarFocus[28]	a-Si-Ge stack	1SC1	107.3 (70.0 active)	6.4 [3]
Solaronix [29]	dye sensitized	MIMO 1010M11	130.0	+
Solarprint [30]	dye sensitized	SP5848	27.8	+
Illumination 200 Lux, artificial light				
China Solar[31]	SC1025I	a-Si	2.4	$3.1*10^{-3}$ [5]
Sanyo Solar Panasonic[32]	Amorton AM-1411	a-Si84	3.5	$3.4*10^{-3}$ unstabilized [6]
Schott* [21]	ASI 2 Oi 06/025/020 JJF	a-Si	5.0	$3.0*10^{-3}$ [7]
Sinonar[33]	SS8223A_BY11	a-Si	19.0	$5.0*10^{-3}$ [8]
Solem[34]	a-Si	05/035/013	4.6	3.1 [9]

*out of business
1 sunny day, clear sky
2 STC
3 at 1000 Wm^{-2}
4 at Air Mass Faktor 1,5
5 200 Lux
6 200 Lux, white fluorescent lamp
7 200 Lux, white fluorescent lamp OSRAM
8 200 Lux, electric
9 200 Lux, fluorescent lamp
+ Information not available

photodiode with 8.8%. As mentioned above, none of these devices had been designed for IPV applications, so provided an optimization for IPV conditions higher efficiencies seem realistic for these materials. Fig.6 shows the relative change of effiency measured in the study in relation to the band gaps. The energy gaps were assumed to be 1.12 eV for crystalline Silicon, 1.44 eV for CdTe, an effective band gap of 1.62 eV for the dye-sensitized photovoltaic device, 1.7 eV for a: Si: H and flexible devices, 1.88 eV for GaInP, and 1.9 eV for P3HT: PCBM.

A current issue in IPV can be seen from Table 2: Some of the devices are fairly competitive under STC, but fail under low irradiance conditions. Thus, the lack of characterization standards can lead to misguiding assumptions when choosing a cell for an application.

Table 2. Photovoltaic efficiencies measured, sorted by best indoor performance (descending) [15], [35]

Material	Manufacturer	Type	Area [cm^{-2}]	η_{STC}[%]	η_{IPV}[%]
GaInP	ISE	prototype	16.0	14.5	15.8
CdTe	TU Darmstadt	prototype	1.6	9.5	10.9
GaAsP	Hamamatsu	G1126-02	1.0	4.1	8.8
Si	ISE	prototype	2.1	20.0	8.2
Si	IXYS	XOB 17-12x1	1.5a	13.0	7.6
a:Si:H	Schott Solar	ASI 2Oi07/037/ 022JJF	8.3	6.4	7.1
dye	ISE	prototype	74.3b	3.0	6.8
P3HT:PCBM	ISE/FMF	prototype	1.2	2.5	5.5
Si	EPS Soltec	prototype	9.0	10.0	5.3
a:Si:H	Sanyo	AM-1417 9 NSJM	4.6	3.2	4.4
Flexible	Flexcell	Gaia 3.6 A - 0070	32.0	3.0	4.2
P3HT:PCBM	ISE/FMF	prototype	1.2	3.1	3.6
Flexible	Plastecs	SP 1.5-50-3	23.3	0.5	3.3

*a. active (total: 1.54) b. active (total: 99.97)

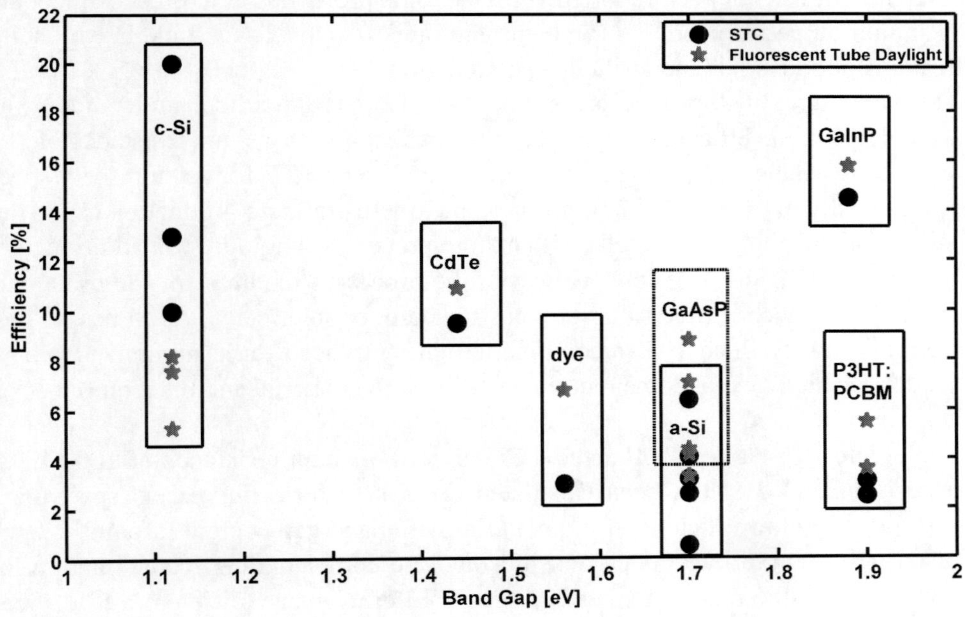

Figure 6. Typical band gaps and efficiencies measured under fluorescent tubes of daylight spectrum (OSRAM Lumilux 840/18 W, grey stars), and under terrestrial solar radiation (AM 1.5, IEC 60904-3, black circles) [15], [35].

4. Future Development

Due to the immaturity of IPV, no photovoltaic cell that has been designed for operation under narrow banded artificial light sources has been demonstrated yet. The ideal band gap of such a cell would be between 1.9-2 eV, depending on the specific light source, resulting in a limiting Shockley Queisser efficiency around 50%. The materials close to these band gap values are P3HT:PCBM and GaInP. From these, GaInP as direct semiconductor with a high shunt resistance is the most promising material in order to obtain high efficiencies. For outdoor applications, GaInP clearly is the more expensive material. The cost for an indoor cell depend on the final design and resulting manufacturing processes. There is also research on the reduction of the rare earth material required for III-V solar cells, mainly by replacing the substrate by low cost semiconductors as for example silicon. The current record efficiency under STC of such a lattice mismatched system is 10% for a GaAs nanowire cell on Si as a substrate [36].

Organic and dye sensitized cells have demonstrated reasonable efficiencies under indoor conditions. In principle, the cost of these printable technologies are low. However, the required minimum lifetime for many indoor applications is in the range of decades, and there a few studies on the longterm performance of organic and dye sensitized cells available.

The second best indoor efficiency of 10.9% has been demonstrated by a chalcogenide material cell, CdTe. The maximum theoretical efficiency for a band gap of 1.45 eV is above 30% for both fluorescent and LED light. For a real cell model of a CIGS cell and a fluorescent spectrum with an intensity of 10 Wm^{-2}, a maximum efficiency of 8% has been calculated (see Fig. 7) [37]. A radiometric indoor characterization of GaAs cell under fluorescent light has not been reported yet. Due to its properties as a direct semiconductor, GaAs should outperform CdTe. The band gap range around 1.4..1.5 eV is well suited to applications of both solar and artificial spectrum.

The maximum Shockley Queisser efficiencies of crystalline silicon under indoor spectra are around 20%. The efficiency of an optimized real cell with a shunt resistance of 100 kΩ and a series resistance of 0.5 Ω has been calculated to be 15% for a fluorescent spectrum with an intensity of 10 Wm^{-2} [37]. In principle, there are large production capacities for crystalline silicon photovoltaic cells with mature process technologies. In addition, the cost of the raw material SiO_2 are low. However, the processes required for the production of reliable indoor devices differ from the ones for outdoor solar cells. As Si has an indirect band gap, Shockley Read Hall recombination currents need to be minimized for indoor applications, which increases the required quality of the material and the required efforts in surface passivation.

The efficiency of a real optimized a-Si cell with a shunt resistance of 100 MΩ and a series resistance of 1 Ω has been calculated to be 14% for a fluorescent spectrum [37]. This considerably low efficiency for a material with band gap energies around 1.7..1.8 eV is caused by the high defect state densities and the corresponding recombination losses. However, besides the reported spread of measured efficiencies, different a-Si devices are available at low cost, and a-Si cells and modules represent the major share of the current IPV market.

It can be seen from this discussion, that the IPV market still shows a large potential for new high efficiency products either by further optimization or by use of new materials. Furthermore, radiometric IPV characterization standards are required for IPV design in research and industry.

Another critical aspect of IPV that has been described in Eq. 5 is its strong angular

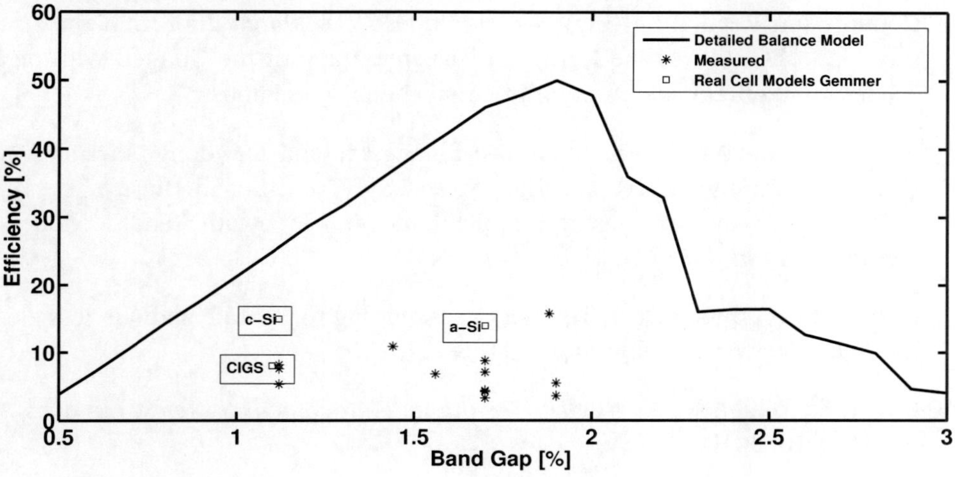

Figure 7. Ideal, simulated, and measured efficiencies under fluorescent tube of daylight spectrum (OSRAM Lumilux 840/18 W) [15], [35], [37].

dependency and the necessarity to operate even under diffuse radiation. An approach to overcome this limitation is to increase the optical efficiency by the use of fluorescent collectors. These devices act as photon traps and spectral converters for radiation of each angle of incidence. The potential of this technique for IPV is still to be exploited both theoretically and experimentally.

IPV has its special challenges due to the typical low irradiance and the often required use of diffuse radiation which still need to be adressed in the design of IPV. Especially because of this and due to the recent change to narrow banded lighting sources, this field of photovoltaics carries great opportunities for the following years.

References

[1] Müller, M., Wienold, J., Walker, W. D., and Reindl, L. M., Characterization of indoor photovoltaic devices and light, *Proc. 34th IEEE Photovoltaic Specialists Conference*, Vols 1-3, New York, 491-496, 2009.

[2] Müller, M., *Energy autarkic micro systems with special focus on indoorphotovoltaics* [German]. Der Andere Verlag, Osnabrück, Germany, 2010.

[3] Roen, S. A., *Solar powered portable calculator*, Litton Business Systems, Inc., Apr. 12, 1977.

[4] Conti, J. A., *Electronic postage weighing scale*, Pitney-Bowes, Inc., Apr. 11, 1978.

[5] Hunt, D. R. G., The Use of Artificial Lighting in Relation to Daylight Levels and occupancy. *Building and Environment* 14(1), 21-33, 1979.

[6] Commission Regulation (EC) No 244/2009 of 18 March 2009 implementing Directive 2005/32/EC of the European Parliament and of the Council with regard to ecodesign requirements for non-directional household lamps.

[7] Hanson, S. and Mingoo Seok and Yu-Shiang Lin and Zhiyoong Foo and Daeyeon Kim and Yoonmyung Lee and Liu, N. and Sylvester, D and Blaauw, D., A Low-Voltage Processor for Sensing Applications With Picowatt Standby Mode, *IEEE Journal of Solid-State Circuits*, 44(4), 1145-1155, 2009.

[8] Paradiso, J.A., and Starner, T., Energy scavenging for mobile and wireless electronics, *IEEE Pervasive Computing*, 4(1), 18-27, 2005.

[9] Beeby, S., White, N., *Energy Harvesting for Autonomous Systems*, London: Artech House, 9-10, 2010.

[10] IEC 60904-3, Ed. 2, *Photovoltaic devices - Part 3: Measurement principles for terrestrial photovoltaic (PV) solar devices with reference spectral irradiance data*, 2008.

[11] Shockley, W., Queisser, H. J., Detailed Balance Limit of Efficiency of p-n Junction Solar Cells, *J. Appl. Phys.*, 32(3), 510-519, 1961.

[12] Freunek (Müller), M., *Indoor Photovoltaics: Efficiencies, Measurements and Design*, In: Tiwari, A., Boukherroub, R., Sharon, M., *Solar Cell Nanotechnology*, Hoboken, NJ, USA: John Wiley & Sons, Inc., 2013.

[13] Glassner, A.(Ed.): *An Introduction to Ray Tracing*. London: Academic Press, II, 1989.

[14] Kurtz, S., Myers, D., McMahon, W. E., Geisz, J., and Steiner, M., A comparison of theoretical efficiencies of multi-junction concentrator solar cells, *Progress in Photovoltaics: Research and Applications*, 16(6), 537-546, 2008.

[15] Freunek (Müller), M., Freunek, M., Reindl, L. M., Maximum efficiencies of indoor photovoltaic devices *IEEE Journal of Photovoltaics*, 3(1), 59-64, 2012.

[16] Prince, M. B., Silicon Solar Energy Converters, *Journal of Applied Physics*, 26, 534-540, 1955.

[17] Hovel, H. J., *Solar cells*, New York: Academic Press, 174, 1975.

[18] Osterwald, C. R., et al., The world photovoltaic scale: an international reference cell calibration program, *Progress in Photovoltaics: Research and Applications*, 7(4), 287-297, 1999.

[19] MÃ$\frac{1}{4}$ller, M., Hildebrandt, H., Walker, W.D., Reindl, L.M., Simulations and Measurements for Indoor Photovoltaic Devices, *Proc. 24th European Photovoltaic Solar Energy Conference*, Hamburg, Germany, 2009.

[20] Hermle, M., Dicker, J., Warta, W., Glunz, S. W., Willeke, G., *Analysis of edge recombination for high-efficiency solar cells at low illumination densities*, Wcpec-3 Organizing Committee, Tokyo, 2003.

[21] Kreitmair, M., Reindl, L. M., Müller, M., Freunek, M., Ungan, T., Micro-energy Harvesting - Current Status of Technology, Field of Applications, and Outlook, *Tutorial Z, Boston MRS Fall Meeting*, 2009.

[22] http://www.eps-soltec.com, accessed 02/20/2014.

[23] http://www.trevesgroup.com/highflex/, accessed 02/20/2014.

[24] http://www.ixys.com, accessed 02/20/2014.

[25] http://www.plastecs.com, accessed 02/20/2014.

[26] http://www.powerfilmsolar.com, accessed 02/20/2014.

[27] http://www.sharp-world.com, accessed 02/20/2014.

[28] http://www.solarmio.com, accessed 02/20/2014.

[29] http://www.solaronix.com, accessed 02/20/2014.

[30] http://www.solarprint.ie, accessed 02/20/2014.

[31] http://www.solars-china.com, accessed 02/20/2014.

[32] http://panasonic.net, accessed 02/20/2014.

[33] http://www.sinonarsolar.com, accessed 02/20/2014.

[34] www.solems.com, accessed 02/20/2014.

[35] Freunek (Müller), M., Reindl, L. M., *Efficiencies, Characterization, and degradation of indoorphotovoltaic converters* [German], VDE Proc. Mikrosystemtechnik Kongress 2013, Aachen, Germany, 2013.

[36] Holm, J. V., Joergensen, H. I., Krogstrup, P., Nygard, J., Liu, H., Aagesen, M., Surface-passivated GaAsP single-nanowire solar cells exceeding 10% efficiency grown on silicon, *Nature Communications*, 4, 1498, 2013.

[37] Gemmer, C. E. M.:*Analytical and numerical studies of solar cellss under different illumination conditions* [German], Phd Thesis, University of Stuttgart, Germany, Osnabrück, Germany: Der Andere Verlag, 2003.

In: Photovoltaics
Editor: Monique A. Gill, pp. 157-171
ISBN: 978-1-63117-843-6

Chapter 8

IRON PYRITE AS A SUSTAINABLE MATERIAL IN THIN FILM SOLAR CELLS

Redhouane Henda*
School of Engineering, Laurentian University,
Sudbury Ontario, Canada

Abstract

The development of sustainable materials in thin film form is a prerequisite for their potential deployment in future photovoltaic technologies. While cost, performance, and recently carbon footprint have been essential factors in materials selection, other equally important factors such as materials abundance (supply chain), chemical risk (toxicity), and recycling/reuse at end of life (cradle-to-cradle approach) have emerged as the new paradigm shift in the deployment of sustainable alternative energy technologies. Many concerted research efforts have been directed at the development of thin films that are able to guarantee optimal characteristics in terms of abundance, environmental compatibility, and photoactivity. Iron disulfide (pyrite) stands out as an excellent candidate for competitive photovoltaic technologies, as it meets all sustainability criteria in addition to its attractive technical capabilities. For over three decades, several investigations have been carried out to develop pyrite in thin film form and highlighted the potential of pyrite in future large-scale production of solar cells. This chapter focuses on the review and discussion of different fabrication methods of pyrite thin films.

The Case for Iron Pyrite

The mineral iron pyrite, formally known as iron disulfide (FeS_2) and hereafter referred to as pyrite, belongs to the transition metal dichalcogenides, MX_2 (where M = Fe, Hf, Mo, Nb, Ru, Ti, W, Zr, and X = S, Se, Te) [1]. It is the most abundant naturally occurring of the sulfide minerals and consists of inexpensive and nontoxic elements (world production in 2011: Fe = 2,940,000,000 tons, unit value = $101.23/ton; S = 70,500,000 tons, unit value = $124.00/ton) [2]. Pyrite crystallizes in a cubic cell structure (similar to NaCl structure) with 12 atoms per unit cell. The latter consists of two interpenetrating fcc lattices (one occupied by

* E-mail address: rhenda@laurentian.ca (Corresponding author)

Fe^{2+} sites and the other occupied by a sulfur-to-sulfur dumbbell), and belongs to the space group symmetry T_h^6 [3]. Pyrite crystal structure is depicted in Figure 1 [4]. FeS_2 can crystallize not only into a cubic structure but also into an orthorhombic (marcasite) structure, which is not photoactive because of its low bandgap energy (0.34 eV). Iron sulfides other than FeS_2, the so-called pyrrhotites ($Fe_{1-x}S$,x = 0-0.17), are predominant in nature, and have different properties and structures depending on the magnitude of the atomic S/Fe ratio [5]. Troilite (FeS), with metal-like conductivity (bandgap of 0.04 eV), is a very common phase whose presence can dramatically reduce film photoactivity. Since both marcasite and troilite regularly coexist with pyrite, it is of utmost importance to devise a preparation route whereby pure pyrite can be obtained. According to some studies [6-8], the stoichiometric range for obtaining pure pyrite is very narrow, i.e., $x \leq 0.5\%$ in $FeS_{2\pm x}$. This can be appreciated from the phase diagram of the iron-sulfur binary system illustrated in Figure 2 [9]. The latter shows a very narrow domain wherein stable pyrite can be obtained without any potential adverse effects from the surrounding non-photoactive phases.

The properties of pyrite are very interesting, especially its semiconducting, electrical, and optical properties [10-13]. A non-exhaustive list is given in Table 1 [8, 14-19]. Of special interest to photovoltaic (PV) technologies are the high absorption coefficient (510^5 cm^{-1}, $\lambda <$ 750 nm), suitable indirect bandgap (0.95 eV), and lattice parameter (5.416 Å) of pyrite. Accordingly, pyrite has many advantages relatively to silicon. Pyrite high absorption makes it suitable for thin film devices, namely, for a 90% light absorption it takes less than 100 nm of pyrite versus 100 μm of Si. Its bandgap is close to the maximum theoretical efficiency limit - highest efficiencies, close to 30% for a single junction solar cell, can be achieved within the band gap range of 1 eV - 1.5 eV – as predicted by Shockley and Queisser [20]. The lattice constant of pyrite differs by 0.3% only from that of silicon (5.431 Å), which is crucial for pyrite films based on silicon substrates. Despite of the favorable properties and apparent sustainability of pyrite, the conversion efficiency of pyrite solar cells has not exceeded 3%. The reason for this low performance is still not yet fully understood [21], but has already been attributed to bulk defects resulting from sulfur deficiency in the films [22], pinning of the Fermi level caused by surface states [23], and, more recently, to coexistence of other phases with pyrite [24]. Due to the optimal properties of pyrite for solar energy conversion, its environmental compatibility, abundance, and low cost, intensive research effort has been devoted to the preparation of highly pure pyrite thin films, for more than three decades (since the first attempt in 1979).

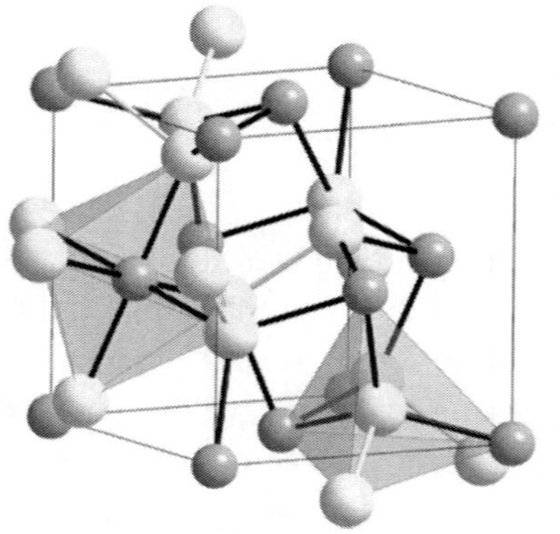

Figure 1. The cubic crystal structure of pyrite [4]. The smaller spheres represent Fe.

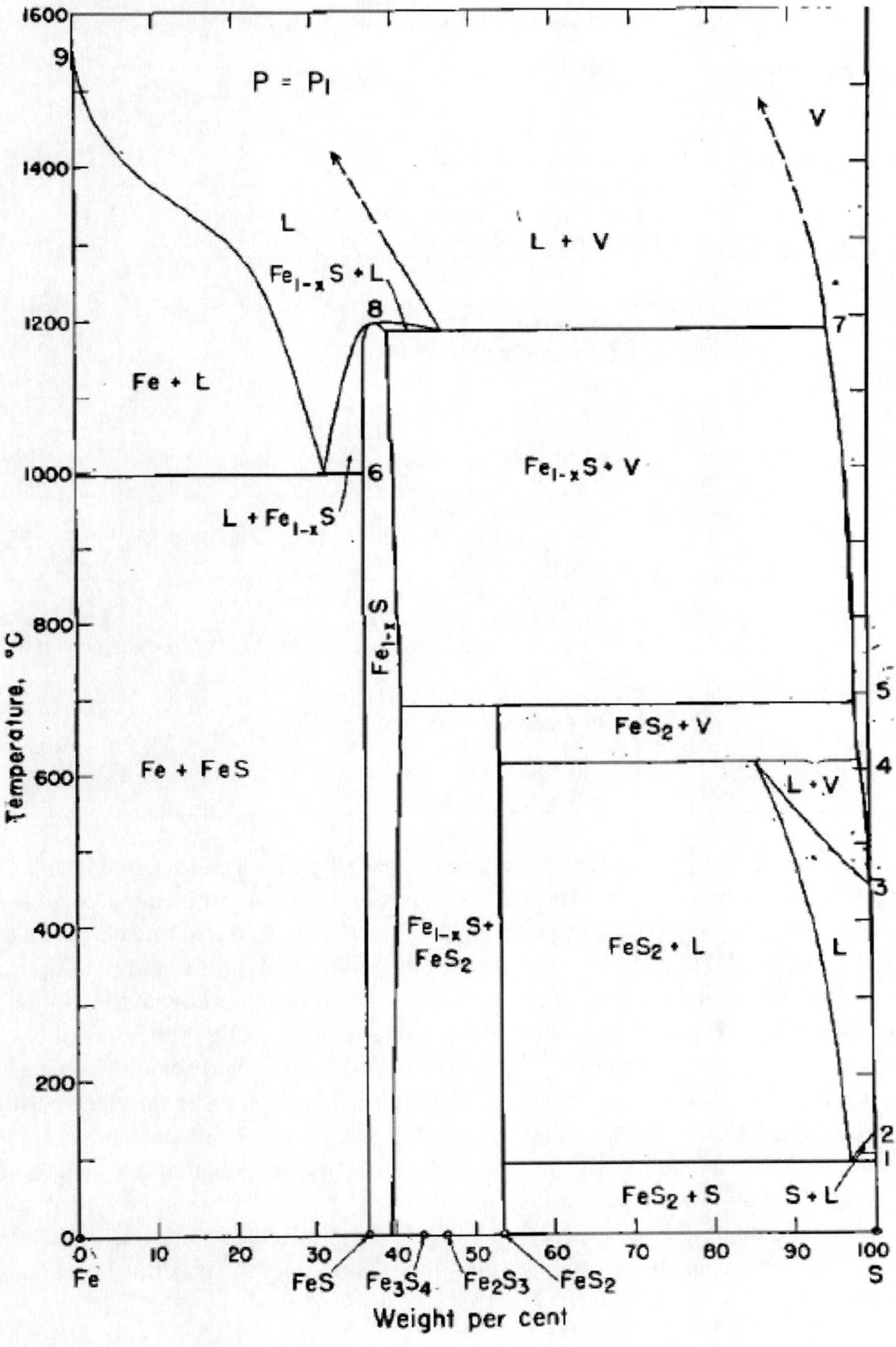

Figure 2. Phase diagram of Fe-S system [9].

Table 1. Properties of Pyrite

Basic properties at 300 K	Atom concentration. $7.5510^{22}/cm^3$ Lattice constant. 5.416 Å Debye temperature. 610^6 K Density. 5.0159 g/cm^3 Dielectric constant. $\varepsilon = 10.9$ Effective electron density. $N_c = 310^{18}$ cm^{-3} Effective hole density. $N_v = 310^{19} \pm 510^{19}$ cm^{-3}
Electrical properties	Bandgap energy (indirect). 0.95 eV Intrinsic carrier concentration. 2.7810^{12} cm^{-3} Carrier mobility. 120 cm^2 V^{-1} s^{-1} Intrinsic resistivity. 0.18 Ω·cm Mobility parameters. $\mu_{max} = 300$ cm^2 V^{-1} s^{-1} $\mu_{min} = 0.02$ cm^2 V^{-1} s^{-1} $c_{ref} = 610^{17}$ cm^{-3} $\beta = 1.3$
Optical properties	Refractive index (average). 4.5 Absorption coefficient. 510^5 cm^{-1} ($\lambda < 750$ nm) Optical phonon energy. 1.048 ± 0.005 eV
Mechanical properties	Bulk modulus. 143 GPa Hardness. 6.3 (on Mohs scale) Surface microhardness (Knoop's pyramid test). 792 kg/mm^2 at 100 Gms load Cleavage planes. (110), (111), (001) Elastic Constants. $C_{11} = 3.46\text{-}3.818$ Mbar $C_{12} = -0.529\text{-}0.34$ Mbar $C_{44} = 0.68\text{-}1.187$ Mbar

A recent study [25] has reiterated the importance of pyrite as a sustainable and potential alternative photovoltaic material. The study in question has demonstrated that there is a major opportunity for the development of large-scale PV technologies based on alternative low-cost and commonly available materials. It is worth mentioning that, in the study in question [25], materials processing and balance of system costs have not been accounted for in the overall cost analysis. These expenditures can make up a significant fraction of the overall cost. Pyrite stands as a frontrunner PV material, as shown in Figure 3, and could achieve between 10^2 and 10^3 as much as the current global annual electricity consumption using the same performance and material assumptions as crystalline silicon [25]. The modest losses in energy efficiency of pyrite, relatively to crystalline silicon, would be offset by the potential for scaling up while saving significantly on materials costs. Moreover, it has been argued that solar cell performance should not be solely assessed against energy efficiency, but also against true cost (related to materials, manufacturing, and installation) and stability as defined by Fonash [26], and expressed as:

$$\text{Cell performance} = \frac{\text{Energy conversion efficiency x Cell lifetime}}{\text{True cost}} \tag{1}$$

Based on this criterion, material thickness becomes a very important feature as it factors in the efficiency and cost of material as well as in material manufacturing. It is clear that highly absorbing, inexpensive, and stable thin film materials are potentially better alternative candidates for optimum solar cells, even though these materials may not have the highest efficiency. Advances in energy efficiency and cost require fundamental science and innovations in fabrication processes. Still, for technology deployment, sustainability factors are key as pointed to earlier, and elsewhere [27].

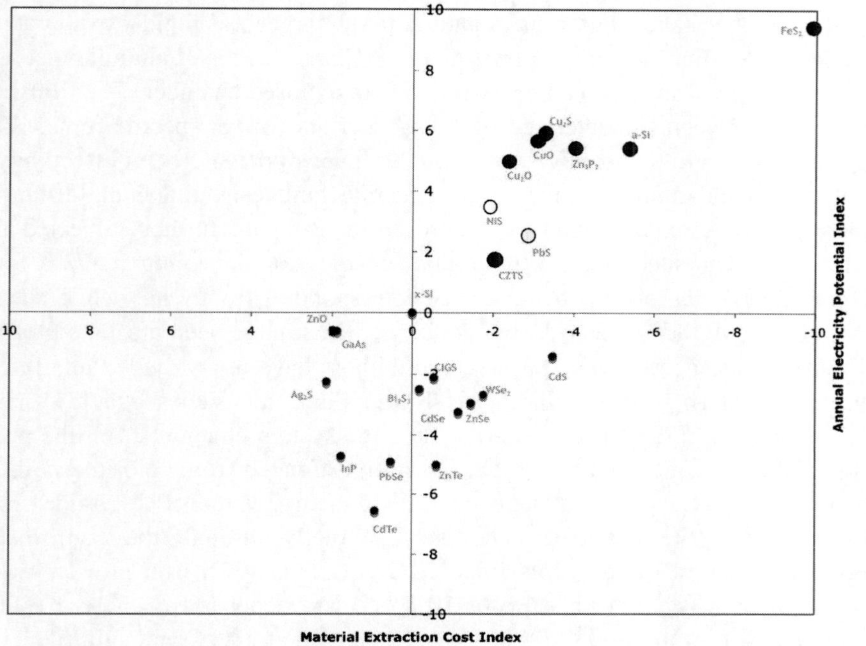

Figure 3. Four-quadrant plot of the annual electricity index versus extraction cost index for the indicated materials. Material extraction cost has been derived from reported mining costs. "Reprinted with permission from [25]. Copyright (2009) American Chemical Society."

It is the intent of this chapter to provide an account of the main experimental techniques for the preparation of thin films of pyrite. Some thoughts on relevant and potential future developments in preparation techniques are also elaborated upon. This chapter does not address patents on pyrite for PV technologies, which have been discussed in a separate contribution [28].

Preparation Techniques

Several thin film deposition techniques have been assessed for the growth of pyrite, over the past three decades. In general, the preparation methods fall under two main categories: vapor-phase based routes and solution-based synthesis routes.

Vapor Phase Routes

Thermal Evaporation

The first attempt to be reported on the preparation of thin films of pyrite is due to Seehra et al. [29] using thermal evaporation. A natural pyrite target was rapidly evaporated, for about 5 min, under a residual pressure of less than 310^{-5}Torr. Glass and aluminum substrates were left in the presence of the evolved gases for 15 h in a closed chamber. The films, of ~300 nm in thickness, have been characterized by Mössbauer and Auger spectroscopies. The analyses were not unequivocally conclusive as to whether pure pyrite has been effectively deposited, and attributed to uncertainties in the measurement techniques. Gupta et al. [30] have followed suit using the same experimental procedure as in [29], but further subjected the samples (~200 nm-thick) deposited on glass and mylar substrates, to annealing at 473 K for 2 h. X-ray diffraction (XRD) has shown three peaks corresponding to pyrite, while the Mössbauer spectrum could only be fitted with two doublets, which have been attributed to the presence of two non-stoichiometric pyrite phases. The authors have not provided any information on the level of deviation from stoichiometry. Karguppikar and Vedeshwar [31] have used the method suggested in [29] but have not isolated the system chamber from the pumping unit. The ratio of atomic concentrations, S/Fe, has been estimated from colorimetry measurements and found to be < 1.5 (in contrast to the stoichiometric value of 2). Aside from this, the identification of pyrite has been attempted indirectly through the confirmation of the semiconducting nature of the films. In a different study [32], iron powder was thermally evaporated under a vacuum pressure of 10^{-5}Torr, and deposited on heated (~200°C) glass substrates. The films, 100-300 nm-thick, were then held at 280°C and sulfurized in a mixture of nitrogen and sulfur source for 30 min. Pyrite peaks have been observed by XRD with a preferred orientation parallel to (200) plane. The authors have estimated a bandgap of 1.31 eV (larger than the reported one for pyrite) and attributed the difference to an amorphous pyrite phase. The other, more fundamental, reason for the observed discrepancy is the fact that the expression used to estimate the bandgap value is not applicable to pyrite [33].

Chemical Vapor Deposition

The first report on the preparation of pyrite films by chemical vapor deposition (CVD) is due to Chatzitheodorou et al. [34]. The films were deposited on glass substrates at near atmospheric pressure (750 Torr) using iron pentacarbonyl, $Fe(CO)_5$, sulfur and hydrogen sulfide, H_2S. Based on XRD and conductivity measurements, polycrystalline pyrite films (0.05-2.00 μm-thick) have been obtained at 130°C. Generally, atmospheric pressure deposition does not result in uniform films, causes unnecessary depletion of the reactants, and is prone to the introduction of impurities in the films. Schleich and Chang [35] have investigated the deposition of pyrite by CVD at low pressure (1.6-5.0 Torr) on various substrates, including glass and quartz. Using $Fe(CO)_5$ and tert-butyl disulfide (TBDS) as precursors, pyrite films have been deposited at temperatures ≥ 480°C. Lower temperatures favor the formation of marcasite. The authors have observed that the use of TBDS, relatively to H_2S, was critical in order to increase sulfur content in the gas phase environment. The high temperature range, while necessary for the growth of pyrite, also results in powder formation, due to undesirable homogeneous (in the gas phase) reactions, and a reduced growth rate of 3 μm/h. Meester et al. [36] have investigated the preparation of pyrite by low pressure CVD (pressure of ~8 Torr) from iron (III) acetylacetonate, $Fe(acac)_3$, TBDS and hydrogen on glass, titanium dioxide and silicon substrates. XRD has shown that the films deposited between 300°C and 340°C consist of polycrystalline pyrite (20-50 nm in thickness) with no detectable to a minor fraction of marcasite in the films. The limiting regime during deposition could not be determined, owing to the complexity of the reaction mechanism, which precludes accurate control over the deposition process. In another study, Takahashi et al. [37] have devised a new CVD system wherein the reaction precursors, $FeCl_3$ and CH_3CSNH_2, are evaporated and carried to a glass substrate using nitrogen gas. The pressure during deposition was held at 1 atm. Pyrite has been obtained at a temperature of 773 K according to XRD data, albeit the atomic ratio S/Fe = 1.94, as per energy dispersive analysis, is below the stoichiometric value implying sulfur-deficient films.

Sputtering

Höpfner et al. [38, 39] have first shown that polycrystalline pyrite films can be produced by reactive ion sputtering in H_2S discharges followed by tempered annealing (with two steps: 100°C and 200°C). Pyrite could not be produced under argon sputtering gas. Deposition was carried out at 75°C and 1.110^{-2} Pa from a synthetic pyrite target on glass and silicon substrates. The crystallinity and stoichiometry of the films have been determined by XRD and Rutherford backscattering (RBS) measurements, which have indicated that the films (48 nm-thick, grain size = 10-15 nm) consist of pure pyrite with an atomic ratio S/Fe of 2±0.05. The study has also pointed to slight deviations of XRD peak reflection positions relatively to the benchmark database. This has been attributed to possible misalignment of the samples during measurements and/or decreased density (found to be 97% of theoretical value, see Table 1) of the produced films. Willeke et al. [40] have reported on the preparation of pyrite films via magnetron sputtering (MS) from a synthetic pyrite target in an argon atmosphere supplemented with a stream of sulfur. The latter was produced by adding a resistively heated, sulfur-filled quartz tube to provide a sulfur atmosphere with argon as the sputtering gas (total pressure: 3.75-22.50 mTorr). Non-reactive MS (in the presence of inert gases only) has led to films with significant deficiency in sulfur content. The authors have observed that substrate temperature, argon flow rate, and the temperature of the sulfur source affect film growth, phase purity, and crystal orientation. XRD analysis has shown that pyrite films (0.40-1.3 μm-thick after 5 h of deposition) are obtained at 350°C, show p-type conductivity and are microcrystalline with an average grain size of about 40 nm. Lichtenberger et al. [41] have applied MS using a mixture of argon and H_2S as the sputtering gas. Two different targets, synthetic pyrite and pure iron, have been used during deposition. Based on XRD and RBS analyses, pyrite films (100-500 nm-thick) have been obtained in a narrow temperature interval around 200°C, for a fraction of H_2S gas of 20%-40%, and a total pressure of 7.5 mTorr.The films have a ratio S/Fe of 1.95 with a substantial amount of oxygen, which the authors have attributed to contamination of H_2S. In a different study, Lichtenberger et al. [42] have increased the pressure of the sputtering gas mixture, i.e., argon and H_2S. The samples have been deposited from an iron target on glass substrates. The pressure of the sputtering gas was increased, relatively to the conditions in [38-41], to within the 7.5-75 mTorr range. This has allowed the deposition temperature to be raised (190°C to 350°C) in the quest to increase film grain size and improve film adhesion. The optimum operating conditions have been pinned down to 80% H_2S gas mixture, total pressure of 37.5 mTorr, and a sputtering power of 100 W. Polycrystalline (grain size = 50 nm) pyrite films have been obtained, as per XRD measurements, with S/Fe ratio of 1.95, according to RBS analysis. The films exhibit p-type conductivity.

Molecular Beam Epitaxy

Bronold et al. [43] have reported on the potential of molecular beam epitaxy in the growth of pyrite films. The samples have been prepared by the evaporation of iron and sulfur from separate molecular beam sources on molybdenum foil and glass substrate. The iron source was heated between 1547°C to 1647°C at ~1.510^{-5}mTorr, and the sulfur source was heated at 977°C at ~510^{-4}mTorr near the substrate. XRD and X-ray photoelectron spectroscopy (XPS) measurements have indicated that polycrystalline pyrite (30-200 nm-thick) can be grown at a substrate temperature of 120°C. Alongside pyrite, the films exhibit FeS and FeS-like phases. The same study has suggested that increasing the sulfur pressure would have improved film stoichiometry, but it is accompanied with a rise in substrate temperature. At higher temperatures (than 120°C), the undesirable phases become predominant or even exclusive in the films.

Thermal Sulfurization of Iron and Iron Oxides

Sulfur deficiency in pyrite films has been a recurring theme. This is the case with compounds that dissociate into constituent atoms, whereby the depositing film upon evaporation of the compound is generally deficient in the gaseous constituent [44]. The loss of the latter can be partially compensated for by replenishing the evaporating environment with a low pressure of the deficient constituent.

Smestad et al. [45] have suggested the sulfurization of iron oxides, Fe_3O_4 or Fe_2O_3, with elemental sulfur as a route to avoid the formation of sulfur-deficient pyrite. Their reasoning is based on the Fe-O-S phase diagram, which shows that reaction pathways of sulfur with the iron oxides in question do not evolve through the FeS phase field. The relevant sulfurization reactions are given as:

$$\begin{aligned} &2\,Fe_2O_3\ (s) + 11\ S \leftrightarrow 4\,FeS_2\ (s) + 3\,SO_2\ (g), \quad \Delta G = -42.8\ \text{kcal/mol} \\ &Fe_3O_4\ (s) + 8\ S \leftrightarrow 3\,FeS_2\ (s) + 2\,SO_2\ (g), \quad \Delta G = -32.5\ \text{kcal/mol} \end{aligned} \tag{2}$$

Reactions (2) can take place in an open system or a closed ampoule. In the present study, sulfur was heated to 350°C for 0.5 to 2 h. Iron oxide layers were prepared in an earlier step via evaporation of iron (1,000 nm-thick) on polyimide and glass substrates, and subsequent oxidation in a mixture of nitrogen and oxygen. XRD and XPS analyses have indicated that the films consist of pure pyrite (grain size: 7-12 nm) with an atomic ratio S/Fe of 2.2 (deviation from 2.0 has been attributed to instrument mal-calibration). Small photocurrents and voltages have been observed in an electrochemical cell, which have been attributed to the high density of grains, presence of pinholes in the films, and the electrolyte solution used to form the electrochemical Schottky barrier was optimized for n-type materials. Hamdadou et al. [46] have obtained pyrite films by the sulfurization of pre-deposited iron layers (80 nm). The latter were produced via electron beam evaporation under vacuum (10^{-5} Pa). Sulfurization was carried out in an ampoule in the presence of sulfur (~10^{-3} Pa) at 823 K for 6 h. The results, from XRD and microprobe analyses, have shown that the films consist of pyrite with S/Fe = 2.08 and grain size of ~100 nm. The purity and composition of the films has been found to be very sensitive to operating conditions, making it difficult to obtain reproducible films. Accordingly, the authors have proposed the use of simpler deposition techniques. In another study by Wan et al. [47], pyrite has been obtained via the sulfurization of iron films (200 nm) in an ampoule at 400°C for 10 min to 48 h. In this case, iron was deposited on silicon by magnetron sputtering at room temperature. Synchrotron radiation XRD data have indicated that better quality (high crystallinity, reduced sulfur deficiency) pyrite films are achieved during long sulfurization times. The study has also delved into the origin of the 'pyrite (200) peak' and concluded that it was due to the silicon substrate via its (400) peak.

Pulsed Power Deposition

Pulsed laser deposition (PLD) and pulsed electron deposition (PED) are relatively new film preparation techniques and based on pulsed energetic bombardment of a target material. In the former case, a pulsed laser beam, and, in the latter case, a pulsed electron beam, rapidly vaporizes a thin spot on the target material, resulting in a stream of energetic plasma flux under highly non-equilibrium conditions. Under optimum process conditions, target material removal, referred to as ablation, can yield congruent film deposition on a substrate. The energy of the pulsed particle (photons or electrons) is in the range of 10–100 eV, well above the common processing temperature of < 0.1 eV (~1200 K). This is quite advantageous when the achievement of high temperature is problematic under equilibrium conditions [48].

Yokoyama et al. [49] have reported the first study on the preparation of pyrite via PLD from a natural pyrite target. The films have been deposited on Al and SiO_2/Si substrates under a vacuum pressure of 7.510^{-7}Torr. The thickness of the films ranges between 610 nm and 1820 nm depending on the substrate temperature (10 K to 633 K). The formation of pyrite has been found to depend significantly on the substrate temperature. At low temperature, Mössbauer results have shown the samples to have a composition close to that of pyrite. The films exhibit sulfur defects and poor adhesion to substrate with less pronounced effects on SiO_2/Si substrate than on Al substrate. At high temperature, observed Mössbauer sextet absorption is indicative of films consisting of troilite. The films also present fewer sulfur defects.

Henda et al. [50] have reported on the pulsed electron deposition (pulse FWHM = ~100 ns) of thin films of pyrite from a synthetic pyrite target. The films (40-120 nm-thick) have been grown on glass substrates in an argon gas atmosphere held at a pressure of 3.510^{-3}Torr, a charging voltage of 15 KV, and pulse frequency of 5 Hz. The substrate temperature was varied between 298 K and 523 K. XPS and visible-Raman spectroscopy as well as XRD analyses have shown the presence of pyrite, mixed with troilite, in the grown samples. The fraction of pyrite has been found to increase with substrate temperature. Recently, ongoing efforts are being devoted to the growth of pyrite via PED from a natural pyrite target. The preliminary results have not been very encouraging, so far [51].

Solution-Based Routes

Electrodeposition

Nakamura and Yamamoto [52] have investigated the electrodeposition of Fe and S from an aqueous solution for the direct preparation of pyrite films. The solution consists of $FeSO_4$ and $Na_2S_2O_3$ (Fe:S = 1:2-100) as source materials. The films have been deposited on titanium sheets for 60 min and under a current density of 0.35-0.75 mA/cm^2. XRD diffractogram has not revealed the presence of any reflection peaks corresponding to pyrite, and XPS measurements have yielded S/Fe ratios close to 1, despite increasing S content of the solution and varying the substrate potential. In order to obtain pyrite, the as-deposited films have been subjected to sulfurization in an open quartz tube for 20 min. The temperature of the films has been varied from 400°C to 600°C. Under these conditions, XRD analysis has revealed that pyrite films can be prepared at 500°C, while XPS data have indicated that S/Fe ranges between 1.9 and 2.2. The goal set up by the authors, viz, to directly electrodeposit pyrite from the co-deposition of Fe and S ions, has proven to be an elusive one. It was further suggested that electrodeposition from a starting solution that contains S_2^{-2} ions would probably result in direct pyrite formation.

Spray Pyrolysis

Abbas et al. [53] have prepared pyrite films for the first time using the spray pyrolysis technique. The reported process involves the spraying of a solution of ferrous salt and sulpho-organic compounds ($FeCl_3$: thiourea: water = 0.15: 0.07: 0.78) onto glass substrates held at 550°C (temperatures < 450° have resulted in films of FeS). Each spraying period consists of 6 s followed by a 5 min-wait period to prevent excessive cooling of the substrate. Subsequently, the reaction produces pyrite films in addition to volatile products. The deposited films (0.5-0.8 μm-thick) have been confirmed to consist of clear, yellow in color, and smooth pyrite, although the study has not mentioned the identification method used. This was elaborated upon in a further investigation as being XRD [54]. Optical absorption data have yielded a direct gap of 1.95 eV and indirect bandgap of 1.12 eV [54]. Smestad et al. [55] have reported on the preparation of pyrite films via spray pyrolysis from the same solution as in [53, 54] but in the molar ratio $FeCl_3$/thiourea of 0.03: 0.072. The solution was sprayed on glass substrates at 350°C under a pressure of 6.3510^{-5}mTorr to 7.6210^{-5}mTorr. XRD data have indicated that the films consist of pyrite with traces of marcasite in their midst. A reaction sequence, to substantiate some of the findings, has been proposed based on the mechanism believed to occur in natural pyrite formation as [56, 57]:

$$\begin{aligned} &(NH_2)_2CS + 2\,H_2O + 2\,HCl \leftrightarrow CO_2 + 2\,NH_3 + 2\,HCl + H_2S \\ &2\,FeCl_3 + H_2S \leftrightarrow 2\,FeCl_2 + S + 2\,HCl \\ &x\,S + H_2S \leftrightarrow 2\,H^+ + S_x^{2-} \\ &FeCl_2 + 2\,H^+ + S_x^{2-} \leftrightarrow FeS_2 + 2\,HCl + (x-2)\,S \end{aligned} \tag{3}$$

The relatively complex reaction mechanism makes it difficult to correlate between film stoichiometry and deposition parameters. In addition to weak photoconductivity, the deposited films exhibit cracks and pinholes.

Thermal Sulfurization of Iron and Iron Oxides

Siyu et al. [58] have reported on the preparation of pyrite from the sulfurization of iron oxide (Fe_2O_3) films obtained through a sol-gel method. A 6 vol.% epoxychloropropane was added to a solution consisting of 20 wt.% ferric chloride and 12 vol.% acetyl acetone in anhydrous ethanol, while stirring and heating at 50°C for 5 h. The resulting solution was clear and brown. The substrate (gold sheet) were dipped into the solution and then pulled into air at a rate of 8 cm/min. Subsequently, the coatings were dried for 15 min in air at room temperature, pre-heated at 100°C for 30 min and heat-treated at 500°C for 30 min in air. The procedure was repeated six times to obtain thicker films. Sulfurization has been carried out as in previous studies, i.e., in an ampoule, at 350°C for 20 h. XRD analysis has confirmed the films (280 nm-thick) to be pure pyrite, albeit energy dispersive spectroscopy has yielded a ratio S/Fe of 1.9 only (no reason has been given for the large discrepancy with the stoichiometry of pyrite). Pimenta et al. [59] have obtained pyrite films from the sulfurization of electrodeposited iron layers using elemental sulfur. Iron thin films were prepared from 0.35 M iron (II) ammonium sulphate solution at 5°C, 20°C, and 80°C, at different current densities, and on different substrates. Sulfurization has been carried out in an ampoule for 30 min. Under all assessed conditions, adhesion of pyrite films was poor, which has been attributed to insufficient cleaning of the substrates, low number of nucleation sites, and/or internal stress in the films. Yamamoto et al. [21] have prepared polycrystalline pyrite on glass and Si substrates by the sulfurization of sprayed films of Fe and S. The authors have used an aqueous solution of reagent-grade $FeSO_4$ and $(NH_4)_2S_x$ as precursors for Fe and S, respectively. Each solution of the precursors has been sprayed alternately onto a substrate held at 120°C. Subsequently, the deposited thin films have been subjected to sulfurization in H_2S flow at a temperature between 400°C and 700°C for 30 min. Pure pyrite films could be obtained at around 500°C using a molar ratio $(NH_4)_2S_x/FeSO_4$ of 2. Diluted $FeSO_4$ solutions (~0.01 M) have yielded enhanced grain growth (gain size of a few μm) and improved electrical properties. All iron pyrite thin films show p-type conduction.

Solvothermal and Colloidal Synthesis

Wadia et al. [60] have studied the preparation of pyrite using a solvothermal route. Materials synthesis, in this case, involves a molecular precursor that undergoes decomposition in the presence of a solvent (surfactant) under moderate to high temperature (100-1000°C) and pressure (between 1 atm and 10,000 atm). In the case of pyrite, an iron precursor and a halide surfactant are used. In this work [60], an aqueous solution of iron (III) diethyl dithiophosphate (in molecular or radical form or a combination of both), as the precursor, and hexadecyl-trimethyl-ammonium halide, $(C_{16}H_{33})N(C_{1-3})_3Br$, as the surfactant, have been used. The precursor and surfactant form an aqueous solution, which is heated to a temperature of at least 150°C and up to 190°C for 2 to 48 h under a pressure in the range between 150 psi and 180 psi. The pH of solution is adjusted (3 and larger) by adding a non-reactive acid (acetic acid) or non-reactive base. The reaction takes place in a stainless steel digestion bomb, and yields a supernatant (murky water) and product (pyrite) that had precipitated to the bottom of the reactor. The precipitate is scraped and rinsed with toluene. XRD, XPS, and energy dispersive spectroscopy have confirmed the structure and purity of the pyrite particles. Transmission electron microscopy has shown the precipitate to consist of aggregated particles (between 100 nm and 500 nm in size). UV-visible absorption spectrum of the particles has resulted in an indirect bandgap value of ~0.90 eV, which matches well with the documented value. Further cyclic voltammetry measurements have reported a bandgap energy of ~1 eV. The particles are insoluble in either water or organic solvents. Addition of an amine (olyelamine) to the mixture of precursor and surfactant has resulted in particles and aggregates that appeared more faceted (and more soluble). The downside is that as the amount of olyelamine is increased, the produced pyrite becomes associated with other phases, viz., marcasite and iron sulphides.

Puthussery et al. [61] have prepared pyrite thin films from colloidal dispersions of single-crystalline nanocrystals of pyrite subjected to sintering in a sulfur atmosphere. The synthesis solution consists of a mixture of $FeCl_2.4H_2O$ (100 mg, 0.5 mmol) and octadecylamine (10 g, 0.14 mol). After degassing at 120°C for 1 h, the temperature of the solution is increased to 220°C under argon gas. An injection solution, prepared by dissolving sulfur (96 mg, 0.5 mmol) in diphenyl ether (5 mL) and degassing at 70°C for 1 h, is added to the synthesis solution. The reacting mixture is held at 220°C and stirred for several hours. The resulting pyrite nanocrystals are separated from the reaction mixture via precipitation in ethanol and purification of the precipitate by washing with chloroform/ethanol mixtures. The nanocrystals are made up of single-phase pyrite without detectable additional phases as confirmed through XRD measurements. UV-visible spectra of the nanocrystals have yielded a bandgap energy of ~ 0.9 eV. Thin films of pyrite nanocrystals are obtained by dipping the substrate (silicon, glass, quartz) repeatedly in the purified nanocrystal solution and a 1 M hydrazine solution in acetonitrile using a dip coater. Subsequently, the films on substrates are sintered in a sulfur-rich atmosphere in the temperature range 500°C - 600°C for several hours, in order to produce high quality, dense, large-grain and stoichiometric pyrite.

Future Prospects

Recurring themes in the different preparation methods discussed so far are film stoichiometry (generally, owing to sulfur deficiency), phase purity of films (due to narrow window of pyrite phase field), and the degree of crystallinity (small grain size). These aspects, in addition to surface/interfacial defects, have an adverse effect on the photoconductivity and stability of pyrite-based solar cells. Based on the intensity of past and ongoing research activity on thin films of pyrite, such efforts do not seem to be waning anytime soon. It seems that pyrite has too many attractive properties to be quickly abandoned as an alternative PV material. The primary motivation for pyrite, and similar materials, revolves around the following sustainability paradigm: Abundance and low cost; appropriate photoactivity; environmental compatibility; recyclability at end of life; and stability (see Figure 4). Lately, according to Fonash's performance criterion, Eq. (1), and sustainability assessments, the prominent view is that highly absorbing, inexpensive, and stable materials may be the best candidates for optimum solar cells, even though these materials may not have the highest energy efficiency. The modest losses in efficiency of solar cells based on such materials would be offset by the potential for scaling up while saving significantly on materials costs. In the case of pyrite, due to its low cost, only 4 % energy efficiency is needed to be competitive with current-generation 20 %-efficient absorbers. This goes without saying that simultaneous achievement of the aforementioned criteria is still a major undertaking, as the present analysis has shown.

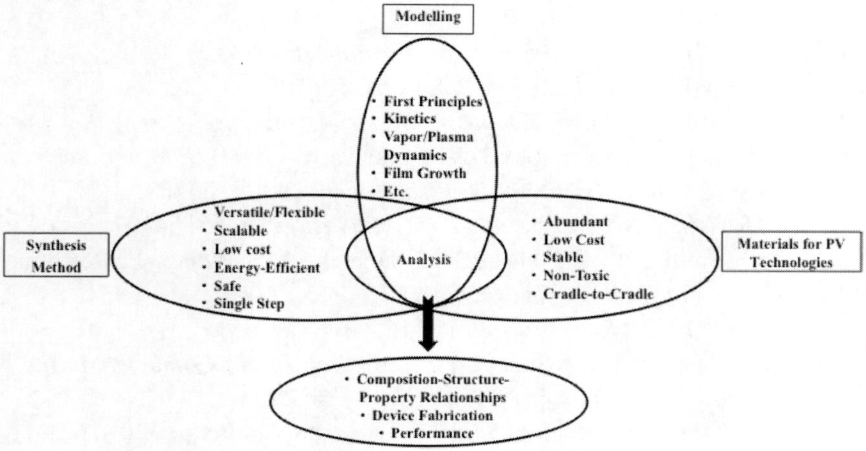

Figure 4. Diagram highlighting a potential integrated approach for the development of materials for PV applications, and the synergies (not shown) between material selection, synthesis, and modeling.

As the quest for alternative sources of energy has been gaining increasing momentum, it is very much likely that research on pyrite thin films for PV applications will continue in the future. It appears that important steps (breakthrough) still need to be undertaken toward improved existing methods and/or the development of new preparation techniques with the proper set of requirements. Modeling efforts should be considered as an integral part in such endeavors, as highlighted in Figure 4. Ideally, a practical process should allow for the direct preparation (preferably in a single step) of pyrite thin films on substrates from a single solid-state source and without potential noxious by-products. It is also critical, as it has been put forward in a recent study [24], to pay close attention to process conditions of each step during device (module) fabrication, so as to prevent potential adverse and irreversible effects on the integrity of active pyrite layers.

References

[1] Wilson, J. A.&Yoff, A. D. (1969). *Adv. Phys.*, 18, 193.

[2] U.S. Geological Survey, Iron and Sulfur statistics, in Kelly, T.D. and Matos, G.R., comps., Historical statistics for mineral and material commodities in the United States: U.S. Geological Survey Data Series 140(2012).

[3] Slate, J. C. (1965). *Quantum Theory of Molecules and Solids*, vol. 2, McGraw-Hill, New York.

[4] Downs, R. T. &Hall-Wallace, M. (2003).*The American Mineralogist Crystal Structure Database. American Mineralogist*, 88, 247.

[5] III Toulmin, P. &Barton, P.B. Jr. (1964). *Geochimocaet Cosmochimica Acta*, 28, 641.

[6] Ellmer, K. &Höpfner, C. (1997).*Philosophical Mag. A*, 75(4), 1129.
[7] Ellmer, K. &Tributsch, H. (March 2000). "Iron Disulfide (Pyrite) as Photovoltaic Material: Problems and Opportunities", *Proc 12th Workshop on Quantum Solar Energy Conversion - (QUANTSOL 2000)*, (Wolkenstein, Italy).
[8] Altermat, P. P., Kiesewetter, T., Ellmer, K. &Tributsch, H. (2002). *Sol. Energy Mater. Sol. Cells*, 71(2), 181.
[9] Kullerud, G. &Yode, H. S. (1959). *Econ. Geology*, 54, 533.
[10] Ennaoui, A. &Tributsch, H. (1984). *Sol. Cells*, 13, 197.
[11] Birkholz, M. (1992). *J. Phys.: Condens. Matter*, 4, 6227
[12] Kato, K., Okamoto, Y., Morimoto, J. &Miyakawa, T. (1997). *J. Mater. Sci. Lett.*, 16, 914.
[13] Barnard, A. S. &Russ, S. P. (2009).*J. Phys. Chem. C*, 113, 5376.
[14] Muscat, J., Russo, S. &Yarovsky, I. (2002). *Phys. Rev. B*, 65, 54107.
[15] Rieder, M., Crellin, J. C., Sustai, O., Drabek, M., Weiss, Z. &Klementova, M. (2007).*Intl. J. Coal Geology*, 71, 115.
[16] Wooste, W. A. (1953). *Rep. Prog. Phys.*, 16, 62.
[17] delasHeras, C., Ferrer, I. J. &Sanchez, C. (1994). *J. Phys.: Condens. Matter*, 6, 10177.
[18] Simmons, G. &Birch, F. (1963). *J. Appl. Phys.*, 34, 2736.
[19] Robertson, F. (1961). *Geological Society of American Bulletin*, 72(4), 621.
[20] Shockley, W. W. &Queisse, H. J. (1961).*J. Appl. Phys.*, 32, 510.
[21] Yamamoto, A., Nakamura, M., Seki, A., Li, E. L., Hashimoto, A. &Nakamura, S. (2003). *Sol. Energy Mater. Sol. Cells*, 75, 451.
[22] Fiechter, S.,Birkholz, M., Hartmann, A., Dulski, P., Giersig, M., Tributsch, H. &Tilley, R. J. D. (1992). *J. Mater. Res.*, 7, 1829.
[23] Ennaoui, A., Fiechter, S., Smestad, G. &Tributsc, H. (1990). "Preparation of iron disulphide and its use for solar energy conversion", Proceedings of 1st World Renewable Energy Congress, 1, Sept. 1990. (Reading, UK) 458-64.
[24] Yu, L., Lany, S., Kykyneshi, R., Jieratum, V., Ravichandran, R., Pelatt, B., Altschul, E., Platt, H. A. S., Wager, J. F., Keszler, D. A. &Zunger, A. (2011). *Adv. Energy Mater.*, 1, 748.
[25] Wadia, C., Alivisatos, A. P. &Kammen, D. M. (2009). *Environ. Sci. Technol.*, 43(6), 2072.
[26] Fonash, S. J. (2010). *Solar cell device physics*, 2nd ed., Academic Press, New York.
[27] Fthenakis. V. (2010). *Renewable and Sustainable Energy Rev.*, 13, 2746.
[28] Henda. R. (2012). *Recent Patents on Chemical Engineering*, 5, 87.
[29] Seehra, S. S., Montano, P. A., Seehra, M. S. &Sen. S. K. (1979). *J. Mater. Sci. Lett.*, 14, 2761.
[30] Gupta, V. P., Chandra, K. &Srivastava. V. K. (1986). *J. Mater. Sci. Lett.*, 5, 165.
[31] Karguppikar, A. M. &Vedeshwar, A. G. (1986). *Phys. Stat. Sol. A*, 95, 717.
[32] Rezig, B., Dahman, H. &Kenzari, M. (1992). *Renew. Energy*, 2(2), 125.
[33] Ferrer, I. J., Nevskaia, D. M., de lasHeras, C. &Sanchez, C. (1990). *Solid State Commun.*, 74, 913.
[34] Chatzitheodorou, G., Fiechter, S., Könenkamp, R.,Kunst, M., Jaegermann, W. &Tributsch. H.(1986). *Mat. Res. Bull.*, 21, 1481.
[35] Schleich, D. M. &Chang, H. S. W. (1991).*J. Cryst. Growth*, 112, 737.
[36] Meester, B., Reijnen, L., Goossens, A. &Schoonman. J. (1999).*J. Phys.*, IV9, Pr8-613.
[37] Takahashi, N., Yatomi, T. &Nakamura, T. (2004).*Sol. State. Sci.*, 6, 1269.

[38] Höpfner,C.,Ennaoui, A., Lichtenberger, D., Birkholz, M., Smestad, G., Fiechter, S. &Tributsch, H. (April 1991). “Thin Film Preparation of FeS2 (Pyrite) by Argon Sputtering, Plasma Reaction, MOCVD and Spray Pyrolysis”, *10th European Photovoltaic Energy Conf.* (Lisbon, Portugal), 8-12.

[39] Birkholz, M., Lichtenberger, D., Höpfner, C. &Fiechter, S. (1992). *Sol. Energy Mater. Sol. Cells*, 27, 243.

[40] Willeke, G., Dasbach, R., Sailer, B. &Bucher, E. (1992). *Thin Solid Films*, 213, 271.

[41] Lichtenberger, D., Ellmer, K., Schieck, R., Fiechter, S. &Tributsch. H. (1994). *Thin Solid Films*, 246, 6.

[42] Lichtenberger, D., Ellmer, K., Schieck, R. &Fiechter, S. (1993). *Appl. Sur. Sci.*, 70/71, 583.

[43] Bronold, M., Kubala, S., Pettenkofer, C. &Jaegermann, W. (1997).*Thin Solid Films*, 304, 178.

[44] Mattox, D. M. (2010). *Physical Vapor Deposition (PVD) Processes*, 2nd Ed., p. 409, Elsevier, Oxford.

[45] Smestad, G., Ennaoui, A., Fiechter, S., Tributsch, H., Hofmann, W.K. &Birkholz. M. (1990). *Sol. Energy Mater.*, 20, 149.

[46] Hamdadou, N., Khelil, A. &Bernède. J. C. (2003).*Mater. Chem. Phys.*, 78, 591.

[47] Wan, D., He, Q., Zhang, L., Jia, Q., Zhang, R., Zhang, H., Wang, B. &Wei, L. (2004). *J. Cryst. Growth*, 268, 222.

[48] Strikovski, M. D., Kim, J. &Kolagani, Solomon S. H. (2010). *Plasma Energetics in Pulsed Laser and Pulsed Electron Deposition*, in Springer Handbook of Crystal Growth, p. 1193, Springer, Heidelberg.

[49] Yokoyama, D., Namiki, K. &Yamada. Y. (2006).*J. Radioanalytical and Nuclear Chemistry*, 268(2), 283.

[50] Henda, R., Al-Shareeda, O., McDonald, A. &Pratt. A. (2012). *Appl. Phys. A.*, 108(4), 967.

[51] Al-Shareeda, O., Henda, R., Pratt, A. &McDonald. A. M. (2013). *Advances in Nano Research*, 1(4), 219.

[52] Nakamura, S. &Yamamoto, A.(2001).*Sol. Energy Mater. Sol. Cells*, 65, 79.

[53] Abass, A. K., Ahmed, Z. A. &Tahir. R. E. (1986).*Phys. Stat. Sol. A*, 97, 243.

[54] Abass, A. K., Ahmed, Z. A. &Tahir. R. E. (1987).*J. Appl. Phys.*, 61(6), 2339.

[55] Smestad, G., Da Silva, A., Tributsch, H., Fiechter, S., Meziani, N. and Birkholz. M. (1989). *Sol. Energy Mater*, 18, 299.

[56] Horwath. R. (1979). *Science*, 203, 49.

[57] Vaughan, D. &Craig. J. (1978). *Mineral Chemistry of Metal Sulfides*, Cambridge University Press, Cambridge.

[58] Siyu, H., Xinyu, L., QingYu, L. &Jun, C. (2009). *J. Alloys and Compounds*, 472, L9.

[59] Pimenta, G., Schröder, V. &Kautek. W. (1991). *Ber. Bunsenge. Phys. Chem.*, 95(11), 1470.

[60] Wadia, C., Wu, Y., Gul, S., Volkman, S. K., Guo, J. &Alivisatos, A. P. (2009). *Chemistry of Materials*, 21(13), 2568.

[61] Puthussery, J., Seefeld, S., Berry, N., Gibbs, M. &Law, M. (2011). *J. Am. Chem. Soc.*, 133(4), 716.

In: Photovoltaics
Editor: Monique A. Gill, pp. 173-196
ISBN: 978-1-63117-843-6

Chapter 9

EFFECT OF URBAN DENSITY ON PV PERFORMANCE

V. S. Y. Cheng and J. C. K. Tong*
Arup, Building Sustainability Group, Hong Kong SAR, China

Abstract

Buildings consume more than 40% of global energy use and are responsible for 33% of global greenhouse gas emissions. Hong Kong, being one of the most densely populated cities in the world, has buildings consuming over 90% of total electricity annually. Out of the various renewable energy sources, solar power using photovoltaic cells is considered the most suitable for wide-scale implementation in a densely populated city with a large number of high rises. To quantify the potential, a methodology was devised to determine the amount of Building Integrated Photovoltaic (BIPV) and Non-BIPV panels that may be installed on the rooftop and façades of selected existing buildings as well as in open space. With a database of building stock being established, a detailed site survey can then be conducted for the sampled buildings in order to determine and verify the feasible areas for PV installation. This methodology can be applied to other similar cities with a large number of high-rise buildings. In addition, the challenges and details of actual projects with different types of BIPV are described. Operating measurements from the projects can then be used for design verification and further practical design considerations. Coupled with practical operating experience, this methodology can be used to evaluate the effect of urban density on photovoltaic (PV) performance.

Introduction

Energy from renewable sources is viewed as one of the important ways to mitigate global climate change. Industrial development has heavily relied on energy generated by fossil fuels. However, powering technological advancement with this energy source generates greenhouse gases (GHG), which mainstream scientists concluded have caused global climate change. As countries undergo rapid urbanization, more high-rise buildings are constructed in cities because of scarcity of land. Buildings consume more than 40% of the world's energy [1].One of the most densely populated cities in the world is Hong Kong, where buildings consume more than 90% of the city's electricity [2] and produce 60% of its GHG [3].

* E-mail address: jimmy.tong@arup.com, Tel: +852-2908-4568 (Corresponding author)

Greater use of renewable energy can alleviate urbanization challenges. There are many different known renewable energy sources at different stages of development. Among them, solar, wind and bioenergy are more mature; but each of them has its own pros and cons. Given the density of high-rise buildings, building-integrated renewable solutions are needed. Whereas utility-scale renewable plants require land space that the city does not have, building-integrated renewable power generation can supply electricity directly to meet the loading demand, thus minimizing the transmission losses. Among mature renewable technologies, solar PV has the longest history as a building-integrated solution.

Building Integrated Photovoltaic (BIPV) panels or arrays are designed to be incorporated on the roofs and façades of buildings, while Non-BIPV systems are used for areas other than building structures, like open spaces, hill slopes, roadside noise barriers and bus shelters. Both BIPV and Non-BIPV can be grid-connected or stand-alone systems. The primary design focus of using solar energy is to delivery power to meet local demand, while passing back excess power to the grid if the system is connected. In addition to the PV system, the overall system also consists of invertor, energy storage, monitoring and metering devices in order to be fully functional.

Given the complexity of cities with high-rise buildings constructed with different purposes and shapes at different periods of time, PV application faces various challenges for new building developments as well as existing building retrofitting. An understanding of this urban effect on PV performance is important to estimate the overall potential for this type of technology. While the cost of implementing solar PV has been reduced due to maturity, the cost is still considered higher than other alternatives. In order to promote wider use of this type of technology, estimating city-wide power generation potential, in addition to the site-specific design situation, will allow governments to develop appropriate policies as well as provide evidence to individual project developers on the potential saving to them and the environment. Therefore, this chapter is devoted to introducing a methodology for assessing PV performance in the urban setting.

This urban PV methodology first involves a study of global solar radiation in the given city or location. Because solar intensity varies in different parts of the world due to different azimuth and inclination angle of the sun, maximum energy generation is limited by global solar radiation. Second, solar cell efficiency should be taken into account, because it is the technological barrier in converting the potential solar energy source to electricity. After considering these two factors, the information is consolidated and analyzed for buildings and open spaces in the city. By applying the selection criteria to eliminate locations with low potential and conducting surveys to verify actual PV applicable feasibility, the city-wide potential can be determined.

To put the different steps of this methodology into context, a case study of Hong Kong is presented. Further to the study of city-wide PV potential, specific project challenges for implementation in individual buildings are summarized. Finally, an actual project experience is documented to uncover the urban effect on the PV system's final performance.

This urban PV methodology first involves a study of global solar radiation in the given city or location. Because solar intensity varies in different parts of the world due to different azimuth and inclination angle of the sun, maximum energy generation is limited by global solar radiation. Second, solar cell efficiency should be taken into account, because it is the technological barrier in converting the potential solar energy source to electricity. After considering these two factors, the information is consolidated and analyzed for buildings and open spaces in the city. By applying the selection criteria to eliminate locations with low potential and conducting surveys to verify actual PV applicable feasibility, the city-wide potential can be determined.

Figure 1. PV in urban city with high-rise buildings.

To put the different steps of this methodology into context, a case study of Hong Kong is presented. Further to the study of city-wide PV potential, specific project challenges for implementation in individual buildings are summarized. Finally, an actual project experience is documented to uncover the urban effect on the PV system's final performance.

Types of Photovoltaic Cells and Efficiency

Since the initial development of the solar photovoltaic technology in 1954, many different types of technology using different material and fabrication process have emerged. Currently, the most common technology for commercial and residential application is the silicon-based solar cell. Three major types of silicon-based solar cells make up about 90% of the solar PV market in the world in 2011 [4].They are widely used in the market for their balance between cost and efficiency.

The three common types of silicon-based solar cell are:

1. Monocrystalline silicon cells

Monocrystalline cells are made out of thin wafers form by single crystals. Typically, the Czochralski process is used, which immerses a crystal nucleus with a defined orientation into a bath of melted silicon and a single cylindrical ingot is drawn very slowly out of the bath. Then, the p-type and n-type areas in the wafer are created by doping with impurities. Generally, the cells are in solid and uniform color. For this type of solar cell made of a higher grade of crystalline silicon, the expected life is longer and most manufacturers offer a 25-year warranty. However, if this type of solar panel is subjected to partial coverage by shade, dirt or snow, the entire circuit can easily break down. The typical efficiency for the monocrystalline silicon cells is 15-27.6%.

2. Polycrystalline silicon cells

Like the monocrystalline cells, polycrystalline cells are also made out of thin wafers, but with multiple crystals. The wafers are cut out from a cast silicon block made out of raw silicon melted and poured into a square mold. As the crystals are formed in various orientations, the surface of the cell looks like shattered glass because of random pattern of crystal borders. Compared to monocrystalline cells, polycrystalline cells are more affected by heat or high temperatures that will lower the performance and shorten the product life. The typical efficiency for the polycrystalline silicon cells is 13-20.4%.

3. Thin-film cells
    a. Amorphous silicon (a-Si)

Amorphous silicon cells are made by applying thin layers of vaporized silicon in a vacuum onto a solid light-weight substrate or flexible backing. The substrate or backing can be made out of glass, plastic or metal. Unlike other crystalline cells, amorphous cells have no distinct crystal structure. Although amorphous cells are not as affected by high temperatures and shading as their counterparts are, their performance degrades more quickly. Typically, the performance can be 15% lower after the cells' initial exposure to sunlight and the drop can take around six weeks. The typical efficiency for the amorphous silicon cells is 5-13.4%.

    b. Cadmium telluride (CdTe)

Cadmium telluride cells are made with cadmium telluride as the p-type area and cadmium sulphide as the n-type area. Compared to other common types of solar cells, cadmium telluride has successfully expanded to multi-kilowatt to gigawatt systems. The typical efficiency for the cadmium telluride cells is 9-19.6%.

c. Copper indium diselenide or Copper indium gallium selenide (CIS/CIGS)

Copper indium gallium cells are made from copper indium diselenide alloyed with gallium and/or sulphur and they use a smaller amount of the toxic material cadmium. Although they are not subject to light-induced degradation, they are still affected by a hot and humid environment. The typical efficiency for the copper indium gallium selenide cells is 10-22.8%.

In addition to the silicon-based solar cell, there are a few special types of solar cells for specific applications, which are designed for higher performance and thus relatively high-cost.

1. Single-junction gallium arsenidecells (GaAs)

Unlike silicon thin film cells, single-junction gallium arsenide cells use direct bandgap semiconductors with high-absorption coefficient material, such as gallium arsenide (GaAs) and gallium-indium phosphide (GaInP) instead of silicon. One drawback of this type of cell is arsenic toxicity. This type of cell is typically found in space application with an efficiency of around 22-29.1%.

2. Multi-junction cells

Most of the afore-mentioned solar cells are made out of a single p-n junction, which can only be designed for a selected narrow energy band or wavelength. Multi-junction cells are formed with multi-layers stacked on top of each other. Each layer/junction is tuned to target a different band gap in order to absorb a wider range of the solar spectrum. Multi-junction cells are formed as lattice-matched, metamorphic and inverted metamorphic with typical material used, such as, gallium arsenide (GaAs), indium phosphide (InP), gallium antimonide (GaSb) and germanium (Ge).The typical efficiency for the multi-junction cells is 31-44.4%.

Furthermore, there are emerging technologies being developed. These technologies have lower efficiency and have not yet become cost-effective, they are designed to have some advantages, in different ways, over existing technologies.

1. Dye-sensitized cells (DSC)

Dye-sensitized cells use semiconductor functions as charge carrier and a light-sensitized organic dye functions as the photon absorber. Semiconductor material like metal oxide (TiO_2, ZnO and Nb_2O_5) is used to coat the cores together with commonly used dye (N3 dye, $RuL_2(NCS)_2$, and Black dye, $RuL'(NCS)_3$)to form the shells. DSCs are quite flexible and semitransparent and believed to be cheap due to its ease of manufacturing. The typical efficiency for the dye-sensitized cells is 11-11.9%.

2. Organic and inorganic cells

Organic photovoltaics received a lot of development for its potentially low-cost production, which uses high-throughput roll-to-roll solution processing. These cells are light, can be fabricated on flexible plastic substrates, and can also be made transparent or color-tunable. These organic semiconducting polymer materials are hole conductors and have an optical band gap that can absorb light in the ultraviolet spectrum.

Beyond using inorganic material silicon, the solar industry continues to explore other abundant, non-toxic, environmentally friendly, sustainable, scalable and inexpensive inorganic material. Cu_2ZnSnS_4 (CZTS), Zn_3P_2 and FeS_2 are viewed as promising materials with large absorption coefficients.

The typical efficiency for the organic and inorganic cells is 5-11.1%.

3. Quantum dot cells

Quantum dot cells use nanoparticles to form semi-conducting crystals and can target different energy bands by controlling the size of the crystal. The cells can be incorporated on conventional substrate such as plastics, glass or metal sheets, or can be molded into a variety of different three-dimensional arrays. The typical efficiency for the quantum dot cells is 4-8.6%.

The efficiency of solar cells is best summarized in Figure 2, which displays the development trends of solar cells. This figure is published by the National Renewable Energy Laboratory [5] every year to record the best solar cell research throughout the world. The major types of cells are shown in groups with different colors. As shown in the figure, multi-junction cells have the highest efficiency, followed by crystalline silicon and thin-film cells, while the emerging technologies are still in an early development stage. With the data provided, a forecasting model was used to predict the "bound efficiency growth" of these technologies [6].It is found that maximum efficiencies for 3-junction, crystalline silicon, CIGS and CdTe cells are 48.5±3.9%, 29.4±2.8%, 25.6±2.4% and 21.4±3.4%.It is estimated that 95% of these maximum efficiency values will be reached around 2020 to 2040, which is shown in Figure 3.

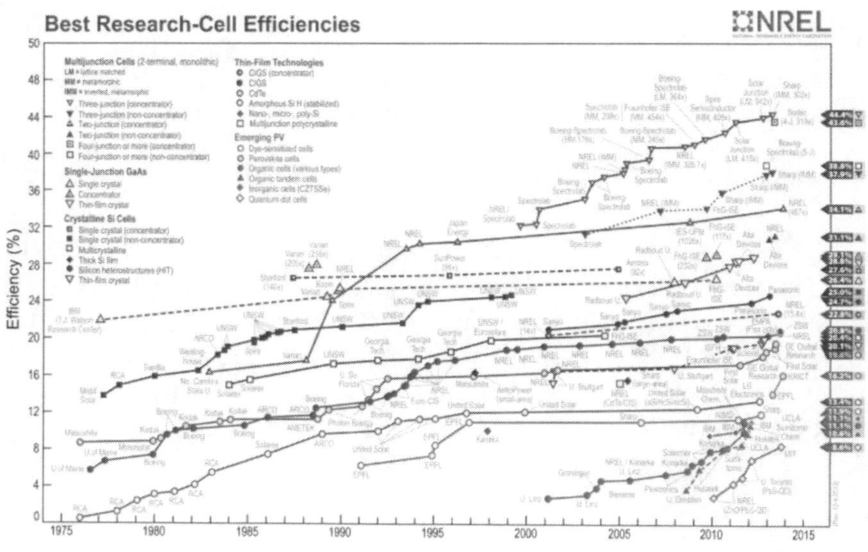

Figure 2. Best Research Cell Efficiencies from 1975-2013 by US National Renewable Energy Laboratory.

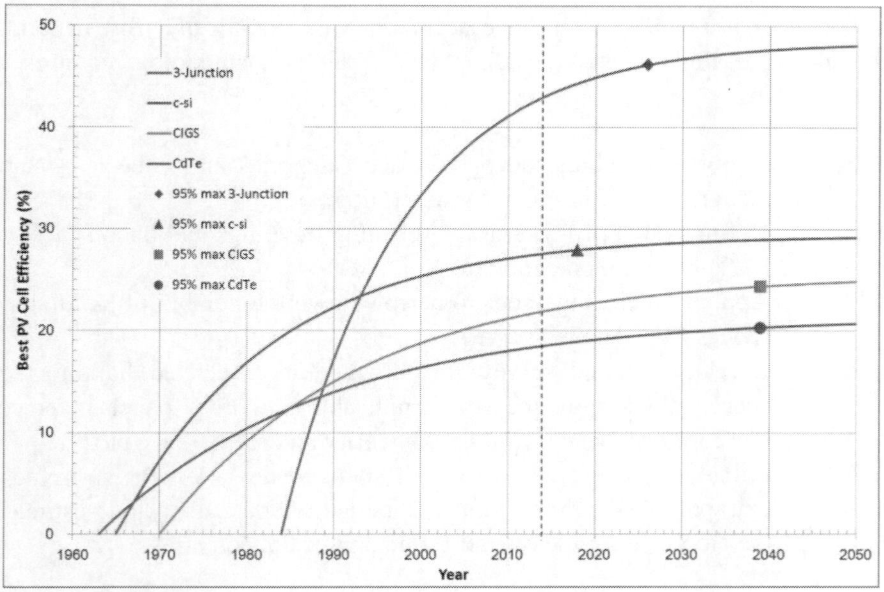

Figure 3. Forecasting from Best Research Cell Efficiencies.

Methodology

To evaluate solar energy potential in the urban environment, not only are local solar radiation and cell efficiency major factors, but also the individual and collective attributes of buildings and open spaces. These attributes are not only important for the evaluation of individual performance, but also the interactive effect of these attributes among buildings and open spaces will affect the overall solar energy potential for an urban situation. For instance, the attributes are orientation, height, footprint area and building age, as well as the shadowing effect from the interaction of these attributes.

Rapid urbanization is taking place in many different parts of the world. Each city faces the challenging task of adopting more renewable energy sources to counter the adverse consequences of carbon emissions. To encourage wider application of solar energy in urban environments with high-rise buildings, a few methods are established to estimate the PV potential. These methods are as follows:

1. Simple empirical method – make use of coefficients based on rule-of-thumb assumptions to take into account the effects of various attributes [7];
2. Photographical method – use aerial photography to estimate the impact of various attributes [8, 9];
3. Geographic Information System (GIS)-based method – use GIS database and light detection and ranging (LiDAR) data to generate 3-D models and use software to evaluate the overall solar irradiation [10, 11].

Each of these methods has its own advantages and disadvantages in terms of accuracy and complexity. In addition to the three methods, this chapter offers yet another methodology.

This methodology is intended to balance accuracy with amount of effort in terms of data collection and processing time and resources. This urban PV methodology comprises with the following five steps:

Step 1. Establish building and open space categories and collect a database with Geographic Information System (GIS);
Step 2. Define and apply criteria for selecting buildings and open spaces with high PV application potential;
Step 3. Apply statistical methods to narrow down the number of buildings and open spaces for a detailed survey;
Step 4. Conduct detailed survey for the randomly selected buildings and open spaces to determine the actual applicable area of PV panels given constraints like roof occupancy, structural integrity and regulations, etc.;
Step 5. Calculate the average area with standard deviation for surveyed samples, extrapolate the results for the entire database and estimate future development for the overall estimation of PV potential.

This urban PV methodology is designed to use statistical techniques to apply the attributes of surveyed set of buildings and open spaces to the overall population of the building stock. It starts with the available data in a large database of buildings and uses different applicable criteria to down-select a smaller list for the actual survey.

Step 1. Establish building and open space categories and collect a database.
The first step is to recognize that the various types of building and open space were constructed for different purposes, and therefore, it is necessary to group them into different categories. For each category, the values of the parameters related to PV installations are within a tighter range. This grouping will enable the subsequent steps of using statistical methods and reducing the sampling size of the survey. The suggested categories and definitions for building and open space are listed in Tables 1 and 2.

Table 1. Categories for building

Category	Definitions
Bus	Business and purpose-built commercial office building
Com	Community center, e.g., library, police station, fire station, hospital, market, and sports complex
Gov	Government office building
Ind	Industrial building
Int	Institution, e.g., primary school, secondary school, university building and educational building
Res	Private and public residential building

To consolidate a database for each category, it is possible to obtain the information about existing building stocks. The required information are the building name, building type, building height and perimeter, surrounding topography, location reference on the map, year built, footprint area, lot boundary and elevation. As for the open space, simply the footprint area is needed. These data can be found in Geographic Information System (GIS)through government agencies and digital maps from commercial means and the public domain.

Table 2. Categories for open space

Category	Definitions
Pk	Country parks, grassland, woodland, farmland, swamp, fishpond and reservoir
Rec	Passive or active recreation, e.g., parks, gardens and football pitches
RR	Roads/Railways, e.g., highway and railway acoustical barriers and enclosures and slopes on the sides of roads or highways

Step 2. Define and apply criteria for selecting buildings and open spaces.

After the long lists of buildings and open spaces for each group are established, it is necessary to identify the relevant criteria for reducing the numbers of locations for a detailed survey. By applying the criteria listed below, shortlists are determined with fewer buildings and open spaces.

Building Criterion 1 (BC1): Age

Typical high-rise buildings are designed for 50 years, however, the life expectancies and functions are different for various groups of buildings. It will not be cost-effective if new PV installation is applied to a building that is nearing the time for demolition. For example, government buildings are expected to have a longer lifespan than the commercial buildings. Therefore, government buildings with fewer than 30 years of age will be considered, as compared to fewer than 20 years for commercial buildings. Moreover, major codes and regulations might alter the PV area availability; hence, the year of implementation should be noted for determining this age factor for each group.

Building Criterion 2 (BC2): Shadowing Effect on the Rooftop of the Object Building Due to Adjacent Buildings

The solar energy generation in buildings is heavily affected by the available annual sunshine hours. The PV panel can only be productive when it is exposed to un-shaded locations. Shadowing effect of a building is determined by the location, orientation, building height and elevation of the rooftop of the building, azimuth and inclination of the sun with respect of different times of year, and the distances from surrounding buildings. As the horizontal surface of rooftops is more effective than the vertical façade, given that the vertical façade is under the shadow while the rooftop is affected by adjacent buildings, the analytical process of determining the shadowing effect is simply considering the rooftop only. Figure 4 shows a schematic cross-sectional picture that illustrates the considered situation. The global insolation availability (GIA) can be determined for each group through this analytical process and the buildings with more than 90% GIA at the rooftop are considered as candidates for a detailed survey.

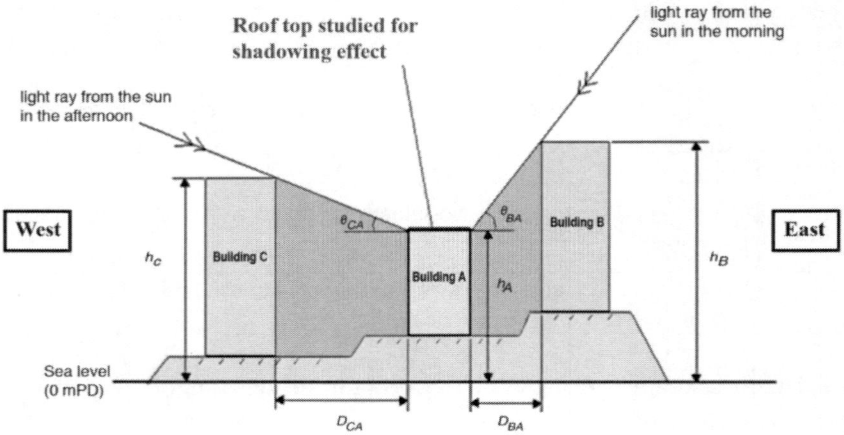

Figure 4. Shadowing effect on building by adjacent buildings with respect to different times of the day.

Building Criterion 3 (BC3): Footprint Area for PV Panel Installation

Another determining factor on solar energy generation is the available space for the PV panel installation. For most buildings, the rooftop area is occupied by chillers, water tanks, gondolas, etc. and the available area is new PV installation will be limited. With the consideration of cost-effectiveness, buildings with at least 10 kW of generation or equivalent of footprint area more than 1,000 m² are considered as candidates for a detailed survey.

Open Space Criterion (OSC): Footprint Area for PVPanel installation

With the consideration of cost-effectiveness, open spaces with at least 100 kW of generation or equivalent of footprint area more than 800 m² are considered as candidates for a detailed survey and by assuming 2% of total area is available for PV installation, an open space with at least 40,000 m²are large enough for the consideration. This assumption is made due to the fact that most area is occupied by trees and plantation and the shadowing effect is not significant.

Step 3. Apply statistical methods for identifying locations for detail survey.

With the buildings and open spaces that are short listed for each group, the number of locations to be selected for the survey is further reduced by using statistical methods for random sampling. As the defined groups are mutually exclusive and collectively exhaustive, stratified random sampling method is used to decrease the variance of the overall sample mean estimate and to study each group individually. Each group or category of buildings and open spaces is considered a stratum with homogeneous attributes, and simple random sampling is applied to each stratum. It is assumed that each stratum has its own means and variances, which can then be used to estimate the mean of the whole population with a reduced variance than from simple random sampling. Stratified random sampling method is widely employed in social opinion and political surveys.

With simple random sampling applies to each stratum, significance testing and null hypothesis are used to verify that a chosen sample size is a good enough to represent the stratum population size.

Table 3. Definition of symbols for statistical analysis [12]

Symbol	Generic Definition	Relevance to this survey	Formula
L	Number of stratum	Name of categories of buildings	-
n	Size of sample population	Total number of all sample buildings for survey	-
n_h	Number of sample elements in a certain stratum 'h'	e.g. Number of sample buildings taken from each category	-
N	Size of entire population	e.g. Total number of buildings	-
N_h	Number of elements in a certain stratum 'h'	e.g. Number of buildings in each category	-
$\overline{X}$	Population mean	e.g. Mean footprint area of all buildings	-
$\overline{X}_h$	Mean of stratum 'h'	e.g. Mean footprint area of buildings in each category	-
$z_{1-(\alpha/2)}$	Coefficient expressing the certainty level of the sample in representing the entire population	Coefficient expressing the certainty level of the sample in representing the entire building population N	Value determined from the Selected Percentiles of Standard Normal Distribution
ε	Percentage error of the mean value obtained from the sampling compared to the true mean of that for the entire population	Percentage error of the mean value obtained from the sampling compared to the true mean of that for the entire building population N	-
π_h	Proportion of sample buildings drawn from a certain stratum 'h' using optimal allocation	Proportion of sample buildings drawn from each building category using optimal allocation	$\pi_h = \frac{n_h}{n} = \left(\frac{N_h \sigma_{hx}}{\sum_{h=1}^{L} N_h \sigma_{hx}} \right)$ (1)
σ_{hx}	Standard deviation of stratum 'h'	Standard deviation of each building category	$\sigma_{hx} = \left[\frac{\sum_{h=1}^{L} \left(X_i - \overline{X}_h \right)^2}{N_h} \right]^{1/2}$ (2)

The following provides further elaboration on the usage of this method – it obtains a greater degree of representativeness to decrease the probable sampling error. This sampling method stratifies the entire building population into several individual subsets called strata, which are mutually exclusive and may be defined according to the choice of stratification variables, which may be building categories, footprint areas, building heights, etc. Within each stratum, the sampling is done independently and a number of elements are chosen randomly. The sample size is determined by the required level of certainty (or confidence level) that the sample represents in the entire population and the percentage difference of the mean value of a certain parameter (e.g. building footprint area) of the sample from that of the entire population.

The definitions of the symbols used for the statistical analysis are summarized in Table 3. The size of the sample n for the entire population to be studied (e.g. total number of all sample buildings for the survey) is given by the equation:

$$n = \frac{\left(\frac{z_{1-(\alpha/2)}^2}{N^2}\right)\left(\sum_{h=1}^{L}\frac{N_h^2\sigma_{hx}^2}{\pi_h \overline{X}^2}\right)}{\varepsilon^2 + \left(\frac{z_{1-(\alpha/2)}^2}{N^2}\right)\left(\sum_{h=1}^{L}\frac{N_h\sigma_{hx}^2}{\overline{X}^2}\right)} \tag{3}$$

The number of samples taken from each stratum is calculated using certain allocation methods. Example of allocation strategies includes: a) equal allocation; b) proportional allocation and c) optimal allocation. Optimal allocation is the best among the three as the standard error is the lowest and thus used. Under this approach, the number of sample elements n_h drawn from each stratum 'h' is determined from the size of sample population n and the standard deviation σ_{hx} of the stratum from Eq.2 as:

i.e.
$$n_h = \left(\frac{N_h\sigma_{hx}}{\sum_{h=1}^{L} N_h\sigma_{hx}}\right)(n) \tag{4}$$

The optimal proportions of sample π_h from each stratum 'h' are first calculated. Afterwards, the total number of sampling elements n (i.e. size of sampling population) is determined by Eq. 3. Different confidence levels (representing different level of certainty for representing the whole population) are tested to show the required samples of buildings for survey.

Step 4. Conduct a detailed survey for the randomly selected buildings and open spaces to determine the actual applicable area of PV panels.

When the exact locations for the buildings and spaces are determined, the survey process is carried out. For each selected existing buildings, the following information is collected:

1. Site plan;
2. Orientation and shadowing effects from the surrounding - including 360 degree photographs of the rooftop;
3. Total footprint and total vertical façades areas;
4. Maximum possible areas of PV panels that can be installed on rooftop and vertical façades;
5. Any other possible areas in the building that can be installed with PV panels;
6. Regulatory and institutional restrictions, including building regulations, height restriction, etc.;
7. Structural integrity for PV installation; and
8. PV support frame conceptual design(s) for integrating into each sampled building.

Figure 5. Typical rooftop of a surveyed high-rise building.

As for selected open spaces, the following information is collected:

1. Site plan;
2. Shadowing effects from the surrounding;
3. Total footprint;
4. Maximum possible area of PV panels that can be installed;
5. Any other possible areas that can be installed with PV panels;
6. Regulatory and institutional restrictions; and
7. Structural integrity for PV installation.

Figure 6. Typical rooftop of a surveyed open space area.

Step 5. Calculate the average area with standard deviation for surveyed samples, extrapolate the results for the entire database and estimate future development for the overall estimation of PV potential.

From the collected data during the last step, the determined potential areas are expressed in percentages. The rooftop PV area is expressed as a percentage of the total footprint of the building. The façade PV area is expressed as a percentage of total vertical façade of the building. The brown PV area is expressed as a percentage of the total footprint of the open space. The mean and standard deviation of the percentages for each category are calculated from the surveyed samples and then applied to the entire shortlist of buildings and open spaces. The following equation is used to convert the uncertainty for each category from the sampled size to the population size of the shortlisted.

$$\bar{x} \pm z_{1-(\alpha-2)} \sqrt{\frac{N-n}{N}} \left(\frac{\sigma}{\sqrt{n}} \right) \quad (5)$$

where $\bar{x}$ is the mean from the surveyed samples

σ is the standard deviation from the surveyed samples

N is the population size in the short list (Step 2.BC3)

n is the sample size in the survey list (Step 3)

$z_{1-(\alpha-2)}$ is the standard normal variable for a particular confident interval (CI) (i.e. 1.645 for 90% and 1.960 for 95%)

With the mean values of the percentages apply to the base areas of the shortlisted buildings and open spaces, the total applicable PV areas are determined. When the global solar radiation for the location is available together with the cell efficiency, the total potential energy generation from solar PV for the existing building and open space is calculated.

The next step is to search for any information for future development. When development projects are planned, further potential can be added to the already estimated values. Even though information will be limited at the planning stage and uncertainty might be high, simple assumptions can be made for the estimation. Therefore, the final estimated potential using this urban PV methodology is available by combining the values for existing buildings and open spaces with the projected values for each type.

Case Study on City-Wide Photovoltaic Potential in Hong Kong

The urban PV potential methodology can be applied in Hong Kong, one of the most densely populated cities in the world. According to the government's annual statistics [2], Hong Kong consumes 90% of its electricity on buildings, and that is equivalent to about 60% of the total annual carbon footprint [3].Hong Kong's main source of electricity is from fossil fuels and only less than 1% is from renewable energy. Hong Kong, with a population of 7.15 million, uses about 25% of land for urban development, out of a total of 1104 km^2 [13].High-rise buildings in the city pose a unique opportunity for applying renewable energy, in order to reduce the reliance on fossil fuels and the associated air pollution. It is important to encourage wider implementation of renewable energy by estimating the generation potential for solar power through this urban PV methodology.

Solar Radiation in Hong Kong

Situated at a latitude of 22°21'N and a longitude of 114°21'E, Hong Kong is located at the southern coastline of mainland China with sub-tropical climate. To estimate solar PV potential, the sun path during different times of the year must be understood. Figure 7 is a sun path diagram for Hong Kong. With consolidated data between 1996 and 2001 from a measuring station set up by the City University of Hong Kong [14], Table 4 shows the annual solar insolation on the horizontal surface, vertical walls facing south, east and west, and an optimally tilted surface at 20°. North facing value is omitted due to the low value in terms of solar potential, which makes the installation cost-ineffective. It was shown that a 4.6% gain can be made with a tilted surface at 20° in Hong Kong [15]. The data has already taken into account seasonal (spring, summer, autumn and winter) as well as daily variations (peak radiation period (11:00-13:00), moderate radiation period (9:00-11:00 and 13:00-15:00) and low radiation period (7:00-9:00 and 15:00-17:00). The presented annual solar insolation values establish the maximum solar energy that can be received at any specific location in Hong Kong.

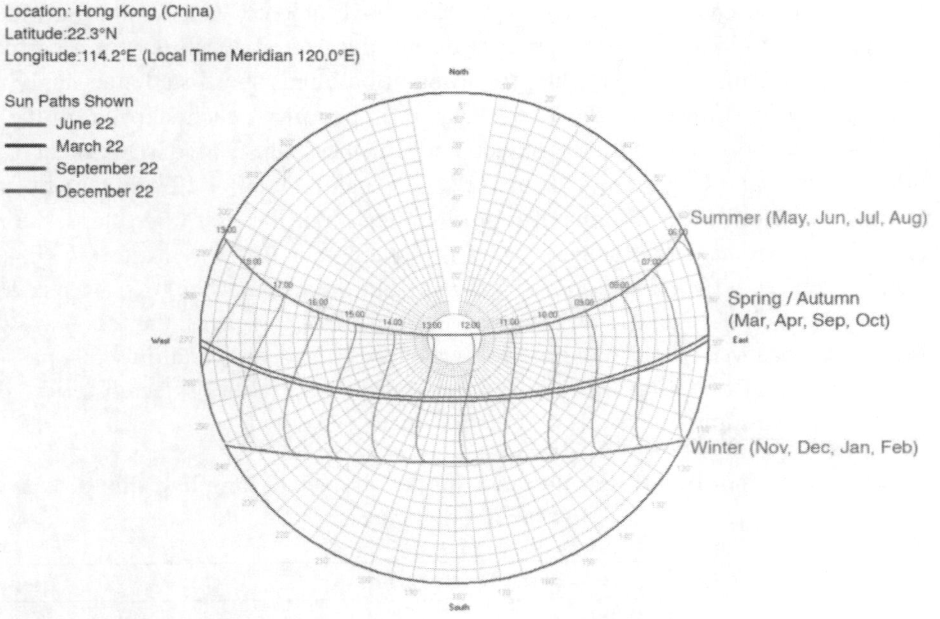

Figure 7. Sun path diagram for Hong Kong with the azimuth and inclination angle of the sun.

Table 4. Annual solar insolation with different orientations in Hong Kong

Orientation	Annual Solar Radiation[kWh/m^2]
Tilted, 20° towards south	1357
Horizontal	1297
Vertical, south wall	752
Vertical, east or west wall	709

Solar PV Efficiency

From the previous section, the efficiency of the common types of PV cells ranges from 10% to 20%.Although actual PV performance is specific to the actual cells used for a project, an average mean efficiency of 15% is assumed for subsequent application of the urban PV methodology.

Summary of the PV Potential in Hong Kong Using the Urban PV Methodology

Using various databases, data for buildings and open spaces of Hong Kong was collected. The focus for estimating the PV potential is only on a subset of the overall population. Since residential and industrial sectors are not as feasible for PV implementation due to multiple ownership and the limited number of locations respectively, this urban PV methodology was used on building categories of Business, Government, Institution and Community Center only. After collecting the necessary information about the buildings, it was found that the distribution for each category is Business (39%), Government (3%), Institution (43%) and Community Center (15%).Table 5 summarizes the number of locations for each category shortlisted using the methodology. The proportion of buildings shortlisted after applying each selection criterion is shown to indicate the relative impact from each step. From the overall long list, the locations are initially reduced to Business (11.3%), Government (32.1%), Institution (11.2%) and Community Center (12.0%); and the total of 12% in terms of number of buildings and about 30% in terms of footprint area. After the first three steps, the number of locations was reduced to Business (0.6%), Government (2.7%), Institution (0.4%) and Community Center (1.1%).Therefore, only 0.7% from the original long list was selected for detail surveying with 90% confidence interval. As for the open spaces, the focus was on Park and Recreation locations. Only one criterion was used to shortlist, and only 5% passed; and using simply random sampling with the combined group, 0.3% was selected for the detailed survey. Table 5 also summarizes this process for open spaces.

Table 5. Proportion of buildings and open spaces on overall long list, shortlist, and final list for detailed survey

	Building					**Open Space**
	Bus	**Gov**	**Int**	**Com**	**Total**	**Pk+Rec**
Long List						
Step 1	100%	100%	100%	100%	100%	100%
Short List						
Step 2.BC1: age	63.3%	81.3%	25.3%	29.0%	42.5%	
Step 2.BC2: GIA 90%	45.4%	69.6%	19.4%	23.2%	31.7%	
Step 2.BC3:>1000m^2	**11.3%**	**32.1%**	**11.2%**	**11.8%**	**12.0%**	
Step 2.OSC: >40,000 m^2						**5.0%**
Survey List						
Step 3: 90% CI	**0.6%**	**2.7%**	**0.4%**	**1.1%**	**0.7%**	**0.3%**

From the sampled locations of buildings and open spaces, the applicable areas for possible PV installation on rooftops and façades for buildings and brown areas for open spaces were identified. Through these representative locations, the mean with statistical uncertainty values of applicable area were calculated and expressed as the percentage of the footprint area of the locations; these values are summarized in Table 6.The values were then applied to the collected footprint areas of shortlisted locations for each category. It is worth noticing that about 5.4% applicable rooftop of the overall footprint area and about 4.1% applicable area of the overall façade area from the long list are considered feasible for PV installation, taking into account the age, shadowing effect and minimal cost-effective area. As for the open space, it is found that more applicable area was available than the estimated 2% to 3.3% from the survey. Together with the solar radiation and cell efficiency with respect to respective orientation, the annual PV generation potential for existing buildings and open spaces are 115.4 GWh and 91.3 GWh, respectively for BIPV and Non-BIPV.

Table 6. PV applicable area in terms of corresponding area from survey findings

	Rooftop[1]	Façade[2]	Brown Area[1]
Gov	3.5% ± 3.3%	13.7% ± 4.3%	
Int	18.1% ± 9.8%	17.8% ± 13.0%	
Com	20.6% ± 9.5%	18.2% ± 9.0%	
Bus	20.7% ± 12.9%	12.3% ± 5.8%	
Pk+Rec			3.3% ± 2.0%

[1] PV applicable area in terms of total footprint area
[2] PV applicable area in terms of total façade area

In addition, a few future development projects were identified. It is difficult to predict, given that very little information is available; however, based on simple assumptions, the estimated PV potential is about 479 GWh and 52 GWh for BIPV and Non-BIPV, respectively.

After exercising the urban PV methodology in a subset of buildings and open spaces in Hong Kong, the total estimated potential energy generated by PV is about 738 GWh annually. This value takes into account various information that is collected and filtered through multiple selection criteria. Furthermore, this value is estimated from statistical means through actual surveyed locations to verify practical aspects for actual PV implementation.

Challenges and Details of Actual Projects

For implementation of PV in the urban setting, many aspects are to be addressed in buildings and open spaces. Challenges in design and construction include regulatory, technical and financial aspects. Law and regulations provide guidelines and limit the possibility of PV installation; technical challenges include applying state-of-the-art techniques and strategies to overcome physical constraints; and finally, the financial aspect drives the pace of development.

Regulatory: Building Regulations

In every local jurisdiction, there are policies, laws, regulations, codes and standards to govern new and existing buildings for BIPV and Non-BIPV applications. Compliance to the applicable regulations is the first consideration for any possible application. For instance, when PV is installed on façade and roof areas, it is necessary to stay within the limits of maximum allowable site coverage and building height. Fire codes normally have stringent requirement to prevent fire from propagating to and from the envelope of the building, therefore, when PV is applied on the outer skin, extra measures on fire safety will be needed. Furthermore, seepage/water-proofing issues on the outer skins of the buildings are to be addressed for code compliance. For PV window application, the building code that governs the minimum lighting and ventilation requirement will limit the size and location for installation. Since many governments are promoting the wider usage of renewable energy, it is possible to get the consent of the development through the authorities even if existing regulations may not fully determine the details of application.

Technical: Physical Constraints

As one of the goals of PV application is to harvest the largest amount of solar energy possible, enhancing efficiency is important. Besides the factors discussed in previous sections, urban air pollution may also significantly degrade the energy production from PV. The air pollution may first block the solar ray from arriving at the surface of the PV panel, and the dust build-up at the surface may also further lower performance.

Another consideration is how the installed PV affects neighboring structures and activities through the light reflection. Neighbors may consider glare as nuisance, and therefore the PV panel position and angle might need to be adjusted to avoid this effect to the surrounding area. Some areas and activities, like the airports, are more sensitive to this effect, and assessment and design modification can be done through reflection/glare calculation and site verification.

Particular to rooftop PV installation on high-rise buildings, strong wind conditions require special attention on structural anchoring. To be code-compliance, the installation of PV panel should be designed for sufficient safety factor and verification testing might be required in order to ensure public safety.

For retrofit projects, the drawbacks (dust, noise, schedule, etc.) of construction work that interfere with normal operations of building occupants should be considered. Minimizing the potential impact will make building owners more willing to install PV panels to existing buildings.

Financial: Market Driver for Implementation

Even though the development of PV has driven the equipment cost down in recent years, the high initial cost is still a barrier of PV implementation compared to other power generation alternatives (both fossil fuel and renewables).As the investment payback is subject to the cost of electricity, equipment and maintenance, and possible subsidies at the time of evaluation, the PV implementation plan should be revisited from time to time.

Another consideration, more frequently found in high-rise residential buildings, is multiple ownership. As more complicated arrangements are required when building ownership is split, it is more difficult to reach a consensus for PV installation.

Operating Measurements from Projects

Following the estimation of applicable solar potential is the final implementation of PV panels in a project. At this point, the building analyzed has gone through the previously described challenges and decided to proceed with the design iteration and installation.

After overcoming the regulatory, technical and financial constraints, the design process was relatively straightforward. The first step was to determine the location within the building premise where space was available, structural integrity was sufficient, and code compliance was in place. Then, the shadowing effect was studied. The design process iterated between the two aforementioned aspects until an optimal solar power generation on a feasible location was identified. The calculation of possible generated power took into account of (1) cell efficiency from the manufacturer, (2) the global solar radiation at the location with shadowing effect with respect to hourly orientation for a year, and (3) network and other efficiencies. Table 7 summarizes the details of the project.

Table 7. Annual PV energy generation from a project

Project Footprint	[m^2]		3305
Cell Area	[m^2]		1015
Architectural Factor			0.31
		Actual	**Design**
Energy	[kWh]	99386	106575
Energy per Area	[kWh/m^2]	97.9	105
Global Solar Radiation	[kWh/m^2]	1288	1311
Supplier Cell Efficiency			14.6%
Shading Efficiency			74.6%
Network Efficiency			75.4%
Other Efficiency			97.5%
System Performance Ratio			0.55
Overall Efficiency		**7.6%**	**8.0%**
Correlation Ratio			**0.95**

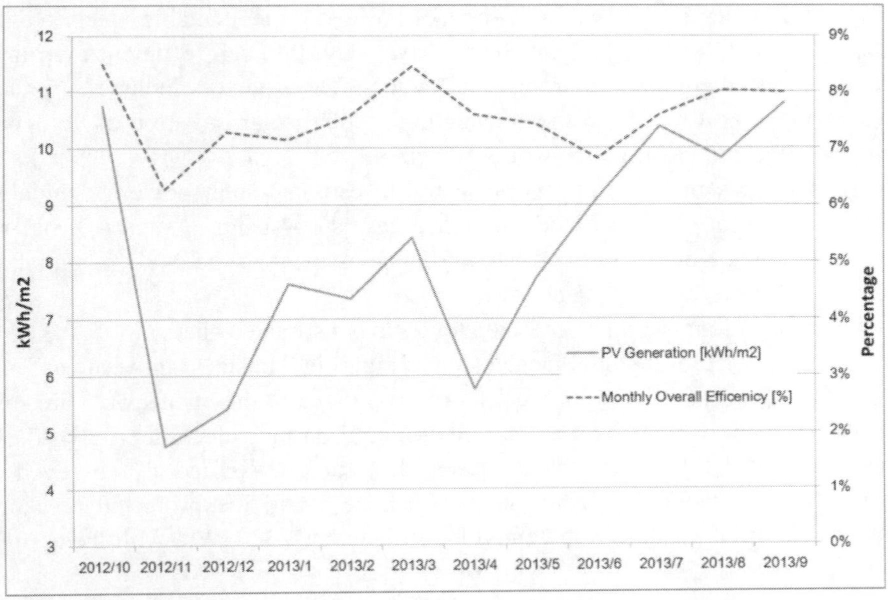

Figure 8. Monthly performance of polycrystalline type of PV cells in Hong Kong.

The project has a footprint of 3305 m^2 and, out of the total area, 1015 m^2 was available for installing polycrystalline PV cells. This area ratio yielded to an architectural factor of 0.31.During the design phase, a cell with 14.6% cell efficiency was employed; and after taking into account of the shading effect and network efficiency, the overall system performance ratio was 0.55.With these design parameters, it was found that an amount of 107 MWh of electricity could be generated, and the deliverable energy was 8% of the solar radiation input. The monthly performance is shown in Figure 8 and it was found that monthly variations of the efficiency were between 6.3% and 8.5% due to many different environmental factors including amount of cloud.

One year of operation was completed and data was collected to find the correlation between the design and the actual value of PV generation. It was found that 99.4 MWh was generated from the PV cells. The difference between this actual value and the corresponding design value was partially because the global solar radiation varied from year to year. With simple calculation, it was determined that the overall efficiency was 7.6%, which is very close to the design value. The 0.4% difference could be due to uncertainties from measurement and other design factors that were unaccounted for.

This correlation effort supports the fact that actual performance matches well with the design performance. Also, this case study provides good evidence for exercising the methodology presented in this chapter. Beyond the 14.6% cell efficiency, the urban factors also contributed further reduction of 6.6%, which is about 45% of the cell efficiency.

Conclusion

In adopting solar energy for buildings and open spaces in urban environments, Building Integrated Photovoltaic (BIPV) and Non-BIPV are the suitable technologies. The performance of PV systems is not only driven by the cell efficiency from the PV manufacturers, but also urban environmental factors. These factors include various building attributes, shadowing effect from the surrounding, global solar radiation of the location and the cloud coverage and electrical network efficiency.

In this chapter, a methodology is presented to estimate solar power potential at a city level. The methodology is illustrated through a case study using a subset of buildings and open spaces in Hong Kong. After completing the assessment of the city's potential, the same approach is used for an actual project.

In the case of Hong Kong at the city level, it is estimated that only 5.4% of the total footprint area and 4.1% of the total façade area of buildings (business, government, institution and community center) and 3.3% of the total footprint area of the open spaces are suitable for PV panel installment. The areas are determined based on the selection criteria like building age, shadowing effect and cost-effectiveness, and further verified by surveyed samples. Together with the existing buildings and open spaces and the projected ones for future development, the total annual PV potential is estimated at 738 GWh, with the assumption of 15% cell efficiency.

Using an actual project to demonstrate, yearly data was collected, analyzed and compared with the design values. The actual performance correlated well with the design estimation, and it was found that beyond 14.6% of the polycrystalline PV cell efficiency, the overall efficiency was 8%.The overall efficiency took into account not only the manufacturer cell efficiency, but also the shadowing and electrical network efficiency.

Through the presented case studies at the city and project levels, it is shown that urban density significantly affects PV performance. Given that each city and project location has its own characteristics, it is important to evaluate the design carefully for cost-effective implementation of solar power.

References

[1] United Nations Environment Programme. (2009). *Buildings and Climate Change: a Summary for Decision-Makers.*

[2] Electrical & Mechanical Services Department (EMSD). (2013). *Hong Kong Energy End-use Data 2013.*HKSAR Government.

[3] Environmental Protection Department (EPD). (2013). *Greenhouse Gas Emissions in Hong Kong by Sector*. HKSAR Government.

[4] The Energy Technology Systems Analysis Programme of International Energy Agency (IEA-ETSAP) and International Renewable Energy Agency (IRENA). (2013). *Solar Photovoltaics Technology Brief* E11.

[5] National Renewable Energy Laboratory. (2013). Best research cell efficiencies. *http://www.nrel.gov/ncpv/images/efficiency_chart.jpg*, accessed January 8, 2014

[6] Morales-Acevedo, A., & Casados-Cruz, G. (2013). Forecasting the Development of Different Solar Cell Technologies. *International Journal of Photoenergy.*

[7] Vardimon, R. (2011). Assessment of the Potential for Distributed Photovoltaic Electricity Production in Israel. *Renewable Energy* (36), pp. 591–594.

[8] Nguyen, H.T., & Pearce, J.M. (2012). Incorporating Shading Losses in Solar Photovoltaic Potential Assessment at the Municipal Scale. *Solar Energy* (86), pp. 1245–1260.

[9] Ordonez, J., Jadraque, E., Alegre, J., & Martinez, G. (2010). Analysis of the Photovoltaic Solar Energy Capacity of Residential Rooftops in Andalusia (Spain). *Renewable and Sustainable Energy Reviews* (14), pp. 2122–2130.

[10] McIntyre, J.H. (2012). Community-scale assessment of rooftop-mounted solar energy potential with meteorological, atlas, and GIS data: a case study of Guelph, Ontario (Canada). *Energy, Sustainability and Society* (2), pp. 1-19.

[11] Byrd, H., Ho, A., Sharp, B., & Kumar-Nair, N. (2013). Measuring the solar potential of a city and its implications for energy policy. *Energy Policy* (61), pp. 944-952.

[12] Levy, P.S., & Lemeshow, S. (1991). *Sampling of populations: methods and applications. (2nd ed).* New York. John Wiley & Sons.

[13] Information Services Department (ISD). (2013). *Hong Kong Yearbook 2012.* Hong Kong SARGovernment.

[14] Li, D.H., Lam, J.C., & Lau, C.C. (2002). A study of solar radiation daylight illuminance and sky luminance data measurements for Hong Kong. *Architectural Science Review* (45), pp. 21-30.

[15] Li, D. H., & Lam, T. N. (2007). Determining the optimum tilt angle and orientation for solar energy collection based on measured solar radiance data. *International Journal of Photoenergy.*

About the Authors

Dr. Vincent Cheng is the Director, Building Sustainability Group, Arup, Hong Kong. He has 20 years of professional experience in building energy efficiency and an industry-known expert in sustainable master planning and building design, specialising in green code formulation, HK-BEAM and LEED assessment, life-cycle analysis and Air Ventilation Assessment (AVA) and District Cooling System (DCS). He has extensive experience on project management for sustainable master planning projects and government consultancy studies, including the formulation of building regulation on Lighting and Ventilation, the CEPAS (BD's code on green building assessment), life cycle assessment tools for EMSD and AVA assessment methodology.

He has also been involved in more than 40 LEED projects in Hong Kong and the region. His completed LEED projects range from Masterplanning (LEED-ND) to commercial Interior (LEED-CI) and has established the design strategies for archiving LEED standards. As a building sustainability specialist, has also been involved in the HK-BEAM projects of Arup and Zero Carbon/Energy Building in the region.

He is the Council member of HKPGBC to advice the government and the professionals on sustainable building issues and also a member in the advisory committee of School of Energy and Environment, the City University of Hong Kong. Internally, he represents HK office for the Arup's Sustainability Task Force and has involved in Arup's development on the strategy for sustainability tools. Dr. Cheng obtained his Ph.D. from the University of Hong Kong.

Selected Publications

[1] Leung, W-H, Yau, R., Cheng, V., Tong, J. (2013). Energy Cascade in Recent Zero Carbon/Energy Development. Paper presented at *Sustainable Building 2013 Hong Kong Regional Conference* selected as 10 Best Papers, Hong Kong, 12-13 September.

[2] Lo, A., Lau, B., Cheng, V. (2013). Challenges of District Cooling System (DCS) Implementation in Hong Kong. Paper presented at *Sustainable Building 2013 Hong Kong Regional Conference* selected as 10 Best Papers, Hong Kong, 12-13 September.

[3] Lam, T., Wan, K., Ng, T., Cheng, V. (2013). Built Environment Modelling and High Performance Envelope Design. Paper presented *at Sustainable Building 2013 Hong Kong Regional Conference*, Hong Kong, 12-13 September.

[4] Au, H., Yan, S.H., Cheng, V. (2013). Modeling Urban Heat Island Effect in a Residential Development in Hong Kong. Paper presented at *Sustainable Building 2013 Hong Kong Regional Conference*, Hong Kong, 12-13 September.

[5] Ng., T.S.K., Yau, M.H.R., Lam, T.N.T., and Cheng, V.S.Y. (2013). Design and Commission a Zero-carbon Building for Hot and Humid Climate. *International Journal of Low-Carbon Technologies*, pp. 1-21.

[6] Wan, K.K.W., Chan, M.H., and Cheng, V.S.Y. (2012). Considerations of Sustainable High-rise Building Design in Different Climate Zones of China. *International Journal of High-Rise Buildings*, pp. 301-310.

[7] Cheng, V. and Wan, K.K.W. (2012). Holistic Considerations for Sustainable Tall Building Design in China. *Proceedings of the CTBUH 9th World Congress*, Shanghai, China, pp.724-731.

[8] Cheng, V.S.Y. and Cheung, G.H.W. (2011). Sustainable Retrofit – the Ultimate Solution for Carbon Reduction in Hong Kong, *Proceedings of the Joint Structural Division Annual Seminar 2011 – Meeting the Challenge of Low Carbon Emission*, Hong Kong, China.

Dr. Jimmy Tong is an Associate at Arup focusing on Building Sustainability. An industrial leader in the energy business, Dr. Tong has applied his expertise in energy systems in various sectors, including wind and renewable energy, infrastructure and building services, and product and system development in the manufacturing of electronics, ventilation equipment, and filtration equipment for more than 15 years. Dr. Tong's current focus is creating and transforming buildings toward a greener future. He obtained a Ph.D. specializing in computational fluid flow and heat transfer from the University of Minnesota.

Prior to joining Arup, Dr. Tong was General Manager at Avantis Limited in Hong Kong leading wind projects around Asia, and has served as a subject-matter expert at Lockheed Martin Maritime Systems and Sensors, Thermal Analysis Partners LLC, Jacobs Civil Inc., Donaldson Company, and XeteX, Inc. in the U.S.

He is also a guest lecturer at several universities in Hong Kong on the subject of energy and sustainability and co-authored book publication and book chapter and publications in archival, refereed journals, including International Journal of Heat and Mass Transfer, Numerical Heat Transfer, Journal of Fluids Engineering, International Journal of Thermal Sciences, Applied Thermal Engineering, Journal of Biomechanical Engineering, and ASHRAE Transactions. Dr. Tong has taught a course in thermal case studies using a commercial software ANSYS/CFX at the University of Minnesota. He also serves as a mentor and project advisor at the American Society of Mechanical Engineering (ASME) and in few universities in Hong Kong.

Selected Publications

[1] Tong, J., Lam, T., Li, T., Chow, K. (2013). Parametric Study for Enhancing Sustainable Building Design. Paper presented at *HK Joint Symposium 2013 "Innovation and Technology for Built Environment",* Hong Kong, 12 November.

[2] Chan, M., Chan, D., Tong, J. (2013). Survival Strategy – Data Mining to Re-energize Tired Assets and Reduce Operating Costs. Paper presented at *HK Joint Symposium 2013 "Innovation and Technology for Built Environment",* Hong Kong, 12 November.

[3] Leung, W-H, Yau, R., Cheng, V., Tong, J. (2013). Energy Cascade in Recent Zero Carbon/Energy Development. Paper presented at *Sustainable Building 2013 Hong Kong Regional Conference* and selected as 10 Best Papers, Hong Kong, 12-13 September.

[4] Yau, R., Tong, J., Ng, T. (2013). Market Drivers on the Transformation of Green Buildings in Hong Kong. Paper presented at *Sustainable Building 2013 Hong Kong Regional Conference* selected as Top 2 Best Papers, Hong Kong, 12-13 September.

[5] Abraham, J.P., Sparrow, E.M., Minkowycz, W.J., Ramazani-rend, R., Tong, J.C.K. (2011). *Modeling Internal Flows by an Extended Menter Transition Model*. In R.J. Marcuso (Ed.), Turbulence: Theory, Types, and Simulation (pp. 149-184). Nova Science Publishers, Inc., Hauppauge, NY.

[6] Abraham, J.P., Sparrow, E.M., Minkowycz, W.J., Ramazani-rend, R., Tong, J.C.K. (2011). *All Fluid-Flow-Regimes Simulation Model for Internal Flows*, Nova Science Publishers, Inc., Hauppauge, NY.

[7] Abraham, J.P., Sparrow, E.M., Tong, J.C.K., Bettenhausen, D.W. (2010). Internal Flows which Transist from Turbulent through Intermittent to Laminar, *International Journal of Thermal Sciences*, v49, pp. 3256-3263.

[8] Sparrow, E.M., Abraham, J.P., Tong, J.C.K. (2004). Archival correlations for average heat transfer coefficients for non-circular and circular cylinders and for spheres in cross-flow. *International Journal of Heat and Mass Transfer* (47), pp. 5285-5296.

In: Photovoltaics
Editor: Monique A. Gill, pp. 197-209
ISBN: 978-1-63117-843-6

Chapter 10

PHOTOVOLTAIC EFFECTS IN NANOSCALE FERROELECTRICS

Jan Seidel*
School of Materials Science and Engineering,
University of New South Wales, Sydney, Australia

Abstract

Light matter interactions in nanoscale ferroelectric materials have received growing interest over the last years due to new developments in scientific instrumentation and novel materials that allow for the study of so far scarcely investigated and /or hidden nanoscale phenomena. Domain walls in complex oxides have especially been in focus opening new pathways for a number of possible applications. [1, 2, 3, 4, 5] Recently an anomalous photovoltaic effect in ferroelectric $BiFeO_3$ thin films was reported that arises from a unique, new mechanism - namely, structurally driven steps of the electrostatic potential at nanometer-scale domain walls. [6, 7, 8, 9] In conventional solid-state photovoltaics, electron–hole pairs are created by light absorption in a semiconductor and separated by the electric field spanning a micrometer-thick depletion region. The maximum voltage these devices can produce is equal to the semiconductor electronic bandgap. Interestingly, domain walls can give rise to a fundamentally different mechanism for photovoltaic charge separation, which operates over a distance of 1–2 nm and produces voltages that are significantly higher than the bandgap. The charge separation happens at previously unobserved nanoscale steps of the electrostatic potential that naturally occur at such ferroelectric domain walls. Their nanoscopic size and flexible arrangement using thin film growth engineering solutions and applied external electric fields offer unique possibilities for novel concepts in oxide based photovoltaics.

Transition metal oxides exhibit a rich physics that can be utilized in a wide variety of properties. Their properties are related to a delicate balance between lattice, charge, spin and orbital degrees of freedom. [10] A large amount of work has been carried out with these materials, especially perovskites, as they offer the advantage of very high structural quality growth. Modern thin film synthesis methods such as pulsed laser deposition (PLD), molecular

* E-mail address: jan.seidel@unsw.edu.au, Website: http://www.materials.unsw.edu.au/staff/jan-seidel.

beam epitaxy (MBE) and metal-organic chemical vapor deposition (MOCVD), among others, enable the engineering of interfaces between complex transition metal oxides with atomic-scale precision. Interfaces introduce a variety of local changes, e.g. they break the symmetry of the material, they induce stress, and they lead to variation in the bonding between ions. This in turn gives rise to changes in electronic structure such as bandwidth, orbital interactions and level degeneracy. Charge transfer in these correlated systems can induce changes in carrier densities at interfaces that result in physical properties which may be completely different from those of the parent bulk phase. [11, 12, 13, 14, 15]

There are many effects of electromagnetic radiation on ferroelectric materials, which are dealt with in textbooks and specialists' works. [16, 17, 18, 19, 20, 21, 22, 23] These phenomena are usually observed and described on the macroscopic and microscopic length scale, where they are more easily accessible. Electrooptic effects and the corresponding constants for example have been recorded for most standard ferroelectric materials, and constitute the basis for current applications. While these properties and applications of ferroelectrics are well explored especially also with regard to their interaction with light, their nanoscale properties upon light-matter interactions now merge into the research focus. A number of interesting associated findings both with respect to fundamental material properties and novel technological applications were reported in recent years, which we will review in this article.

The motivation to study light interaction with nanoscale ferroelectric materials is manifold. Important technological challenges such as for example the worldwide need for alternative energy sources leads researchers to investigate a broad spectrum of new candidate materials for thin film photovoltaic cells hopefully forming an important step towards renewable energy production. [24, 25, 26] The fundamental physics behind state-of-the-art solar cells (for example Si-based solar cells) is charge separation using the potential developed at a classical p-n junction. This begs the question: are there other pathways to accomplish charge separation in materials in order to enable and trigger next-generation photovoltaics? In the past, anomalous photovoltaic effects in polar materials have been found to arise from two mechanisms: (i) as a consequence of granularity [27, 28] and (ii) "bulk" effects arising from non-centrosymmetry. [29, 30] The former mechanism inevitably suffers from poorly-controlled granular interfaces, while the latter may arise only in very good electrical insulators and is typically only seen in materials with bandgaps in excess of 3 eV that consequently absorb only very little of the solar spectrum. Another mechanism of charge separation and photo-voltage (PV) generation was reported to occur at ferroelectric domain walls in $BiFeO_3$ (BFO), under white light illumination. [31] In contrast to semiconductor based photovoltaics, [32] the photo-voltages observed here are significantly higher than the electronic bandgap. In this review we are especially focusing on effects of electromagnetic radiation on ferroelectrics on the nanometer length scale, which provides a new playground for research. In addition, it offers a novel platform for future nanotechnology and oxide based photovoltaics. [33]

Ferroic phases can arise in two or more distinct orientations, therefore they can form domains, separated by domain walls. Domains are a representation of long-range order with respect to at least one macroscopic tensor property of the material (order parameter). Their orientation states can be changed, leading to movement of the domain walls. Many so-called smart materials and structures have at least one form of ferroic order. They can be designed to change this order in a preconceived manner through the application of external fields. Thus, given their intrinsic structure, ferroic materials are ideally suited for functionality on the nanoscale.

Domain walls are naturally occurring interfaces and topological boundaries in complex oxides as opposed to artificially created heterointerfaces between different materials. [34] They have been investigated with respect to the mechanical and leakage current behavior in various studies involving ceramics, thin films and single crystalline material. However, the fact that their electronic conductivity can actually be utilized for nanoscale functional elements has sparked renewed interest and has led to the finding of unique properties associated with such domain walls. First findings in $BiFeO_3$ (BFO) have led to the investigation of other ferroelectrics and multiferroics, which also exhibit electronic conduction at domain walls, albeit in different forms. Similar considerations regarding electronic conductivity apply to structural phase boundaries, [35] which might e.g. occur in materials exhibiting morphotropic phase boundaries, which includes BFO. [36]

Scanning Probe Microscopy (SPM) with its many variations (e.g. conductive Atomic Force Microscopy – c-AFM, Piezoresponse Force Microscopy - PFM) are very well suited for the direct photovoltaic characterization of prototype oxide structures on the nanoscale, including domain walls. [37] With c-AFM one can probe local conductivity at interfaces and topological defects. PFM has been further developed and is now offering dynamic characterization of the switching process utilizing developments such as stroboscopic PFM and PFM spectroscopy. [38, 39, 40, 41] The recently spike of interest in local conductivity measurements at domain walls arises from perspectives involving non-volatile memory applications and a considerable potential for electroresitive memory devices. [42] Such functionality is also interesting for the characterization of oxygen vacancy movement and ionic battery materials on the nanoscale. [43, 44, 45] In many cases, the presence of extended defects and oxygen vacancies makes the identification of polarization mediated transport mechanisms difficult. The combination of local electromechanical and conductivity measurements has been proven as an interesting approach and has revealed a connection between local current and pinning at bicrystal grain boundaries in bismuth ferrite. [46] More generally, electroresistance in ferroelectric structures has recently been reviewed by Watanabe. [47] The presence of extended defects and oxygen vacancy accumulation has also been shown to influence transport and nucleation mechanisms at domain walls. [48, 49, 50]

The emergence of semiconducting ferroelectrics with smaller bandgaps and the possibility of conduction at domain walls, as described below, has stimulated renewed interest in exploring Scanning Tunneling Microscopy (STM) as a probe of local electronic structure in complex oxides. The emergence of combined AFM/STM or SEM/STM systems are especially interesting in terms of exploring the electronic properties of domain walls in such insulating materials. Investigations using such combined tools have been demonstrated and include measurements on BFO domain walls. [51, 52, 53, 54] STM and STS measurements in cross-sectional samples have been used to directly investigate the nature of the local density of states and band structure at ferroelectric domain walls in multiferroic BFO. [31] Samples with engineered stripe domain array were used in order to simplify the location of a single wall in the STM setup for local spectroscopy. It was found that in-situ cleaved samples with ordered stripe arrays show decreases of the bandgap at the domain boundaries (Figure 1). In addition, a shift towards the Fermi level in the band edges of 109 ° and 71 ° domain walls have been observed. The demonstrated approach in this work serves as a model technique to investigate and understand electronic structure at complex oxide interfaces.

The changes in structure (and as a consequence electronic structure) that occur at ferroelectric (multiferroic) domain walls [55] can lead to changes in transport behavior.

Indeed, domain wall conductivity has been shown in different ferroic materials, although with different transport behavior: the domain walls of $BiFeO_3$ were found to be more conductive than the domains [22, 56, 57] while those of $YMnO_3$ were found to be more insulating or conductive depending on their orientation. [58, 59] In multiferroic $YMnO_3$, a so-called improper ferroelectric, in which ferroelectricity is not the primary order parameter but rather induced by structural trimerization coexisting with magnetism, domain walls are found to be charged and stable and their conductivity depends on the specific orientation of the domain wall with respect to the polarization vector.

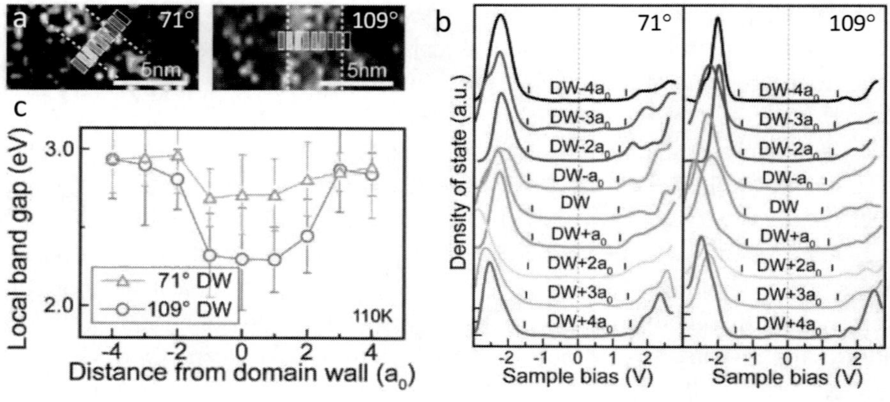

Figure 1. Layer-by-layer d I /d V measurements across 71 ° and 109 ° domain walls in $BiFeO_3$ acquired at 110 K. Bars in a) denote positions where the electronic spectra are probed, and b) show the corresponding STS spectra. The band edges are indicated by black tick marks in b). c) Extracted local band gap across the domain walls. Adapted from Chiu et al [31].

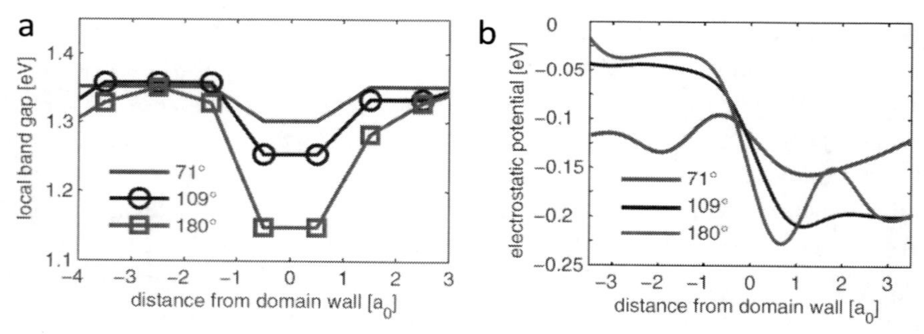

Figure 2. DFT calculations of local band gap at domain walls in BFO extracted from the layer-by-layer densities of states. b) macroscopically and planar-averaged electrostatic potential for the domain boundaries with continuous oxygen octahedral rotations. Adapted from Lubk et al [35].

The electronic structure of domain walls in bismuth ferrite has been investigated by Lubk et al. [60] Using density functional theory (DFT) the layer-by layer densities of states was calculated showing that the domain walls have a significantly reduced band gap compared to the *R*3*c* bulk structure. Structural changes at the walls lead to arrangements that approach the ideal cubic structure, in which the 180° Fe-O-Fe bond angle maximize the Fe 3*d*-O 2*p* hybridization and hence the bandwidth of the material. Figure 2 shows the local band gap extracted from the layer-by-layer densities of states across all three wall types in BFO. In all cases a reduction in the band gap in the wall can be seen, with the 180° wall showing the largest effect. Note that in no case does the band gap approach zero in the wall region. The same first principles calculation supporting experimental work of Seidel et al. [22] They also give insight into the changes in Fe-O-Fe bond angle in BFO. In addition, the rotations of the oxygen octahedra are found to not change their phase across the walls, i.e. this scenario is significantly more favorable than that with rotation discontinuities. The calculations thus show that antiphase boundaries associated with octahedral rotations are energetically very costly.

The same DFT calculations also revealed steps in the electrostatic potential for all domain wall types, which is an important point, as they also contribute to the electronic conductivity. Steps in the electrostatic potential at domain walls are correlated with (and caused by) small changes in the component of the polarization normal to the wall. [21] These changes in normal polarization are a consequence of the specific rotation of the polar vector across the domain wall, and are also found in other materials. Tetragonal $PbTiO_3$, for example, shows a similar effect for a 90° wall. [61] Extended phase-field calculations for tetragonal $BaTiO_3$ also allow calculating the intrinsic electrostatic potential drop across the 90º domain wall, regardless of the consideration of the ferroelectric as n-type semiconductor or dielectric. [62] This potential change creates a large local electric field that promotes an asymmetric charge distribution around the walls, i.e. they foster the accumulation of carriers and oxygen vacancies close to the wall. The increased charge density plays a role in increased conductivity at such walls.

As mentioned before, the semi-rigid rotation of the polar vector across a ferroelectric domain wall leads to an electrostatic potential that is screened by free charges which enhance the local charge density and thus, presumably, the conductivity. Since this polar rotation [63] is not exclusive to BFO, other perovskite ferroelectrics should also be expected to display enhanced conductivity. This increased conductivity indeed has been found for lead zirconium titanate (PZT), [64] hexagonal manganites ($YMnO_3$, $ErMnO_3$) [31, 30] and lithium niobate under illumination. [65] For BFO, several other factors might be further helping the conductivity enhancement: Firstly, the magnetoelectric coupling between polarization and spin lattice is such that the magnetic sub-lattice rotates with the polarization. [66, 67] Since spins rotate rigidly, they might favour a more rigid rotation of the polarization and hence a bigger electrostatic step at the wall (and of course the polarization of BFO is itself bigger than that of other known perovskite ferroelectrics, which means that all other things being equal a rigid polar rotation in BFO will cause a bigger electrostatic step). But perhaps the most obvious consideration is the fact that $BiFeO_3$ has intrinsically smaller bandgap than other prototype perovskite ferroelectrics (~2.7 eV instead of 3.5-4eV). This means that the screening charges accummulated at the wall will be closer to the bottom of the conduction band and hence will more easily contribute to the conductivity.

As an additional factor, the role of defect accumulation at the domain walls also deserves close scrutiny. Localized states are known to appear in the spectrum of ferroelectric semiconductors, and states localized at the walls and inside the domain but close to the wall split off from the bulk continuum. These non-degenerate states typically have a high dispersion, in contrast with the "heavy-fermion" states at an isolated domain wall. [68] This behaviour can lead to charged double layers that can be formed due to coupling between polarization and space charges at ferroelectric-ferroelastic domain walls. [69] Charged domain wall energies are about one order larger than the uncharged domain wall energies, [70] and phenomenological calculations show decoration of walls by defects such as oxygen vacancies, which are always present in oxides. The presence of charge and defect layers at the walls means that such walls can cause electrical failure in ferroelectric capacitor structures by providing a high conductivity pathway from electrode to electrode. [43]

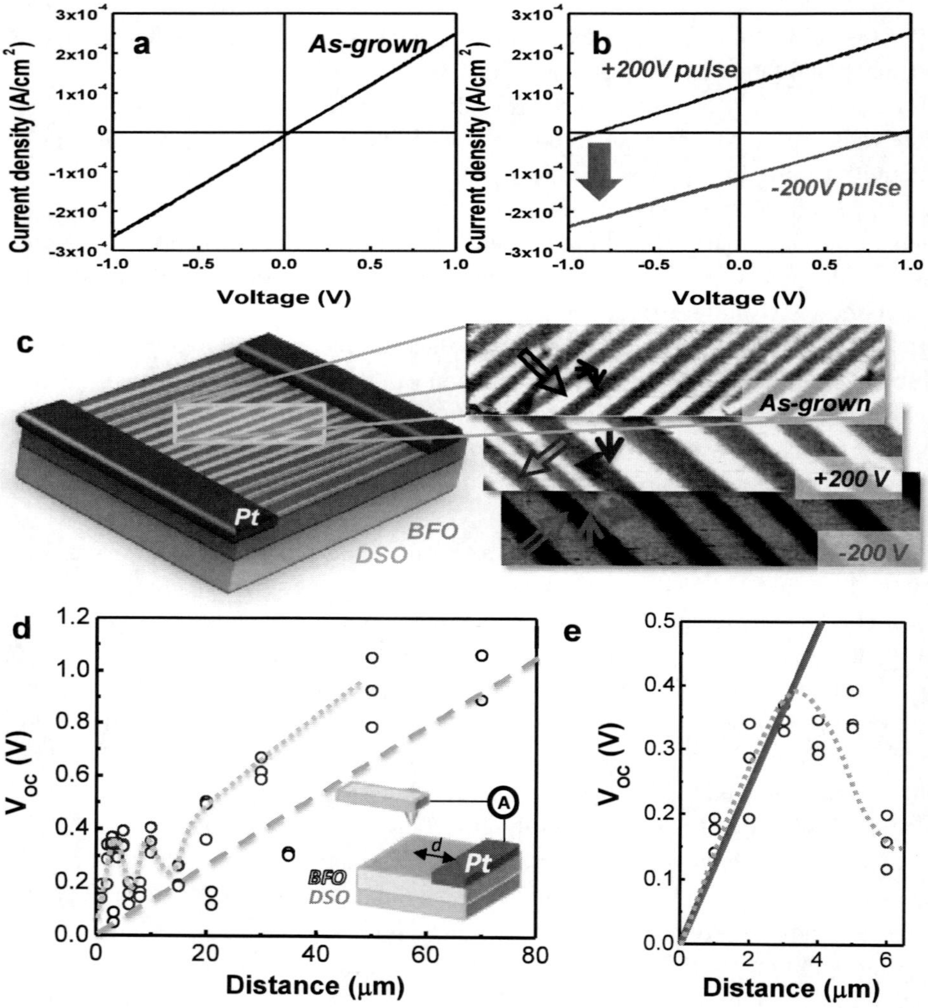

Figure 3. Switchable domain-wall PV effect and c-AFM based PV characterization. a) Photovoltaic I-V curve for as-grown sample, b) Same after switching. c) Corresponding PFM images of the as-grown and poled device structures. Arrows indicate the in-plane projection of the polarization and the net polarization direction. d) Local measurements of V_{OC}. a) 109° degree domain walls (blue) show different, oscillating behavior of V_{oc} with distance as compared to 71° domain walls (black) due to domain superstructure. inset: schematic nanoscale PV measurement setup, b) Initial large slope (red line) indicating large PV effect at 109 degree walls. For details see [16, 48].

BFO exhibits an anomalous photovoltaic effect in thin films arising from a unique mechanism -- namely, structurally driven steps of the electrostatic potential at nanometer-scale domain walls, see Figure 3. [71, 72, 73] In conventional solid-state photovoltaics, electron–hole pairs are created by light absorption in a semiconductor and separated by the electric field spanning a micrometer-thick depletion region. The maximum voltage these devices can produce is equal to the semiconductor electronic bandgap. Interestingly, domain walls can give rise to a fundamentally different mechanism for photovoltaic charge separation, which operates over a distance of 1–2 nm and produces voltages that are significantly higher than the bandgap. Recent investigations using conductive AFM under light illumination reveal these high photovoltages at 71° and 109° domain walls in BFO (see Figure 4). [49] The charge separation happens at previously unobserved nanoscale steps of the electrostatic potential that naturally occur at ferroelectric domain walls in the complex oxide $BiFeO_3$. Electric-field control over domain structure allows the photovoltaic effect to be reversed in polarity or turned off.

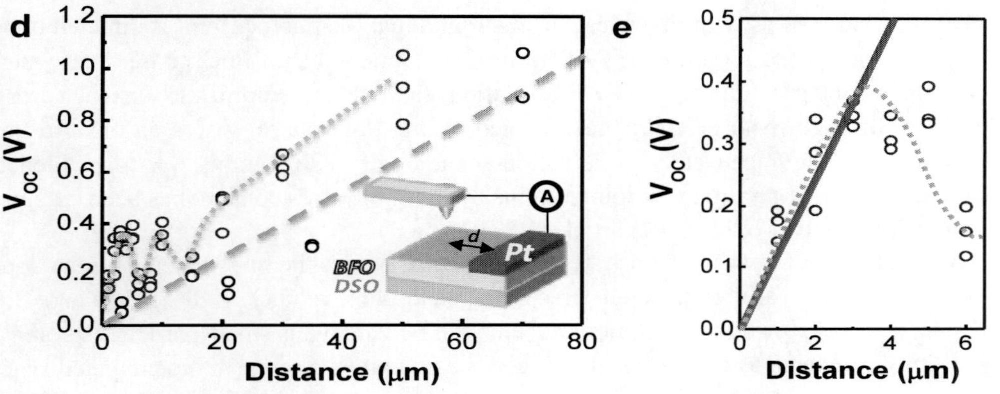

Figure 4. a) Local measurements of open circuit voltage V_{OC} at domain walls in BFO a) 109° degree domain walls (blue) show different, oscillating behavior of V_{oc} with distance as compared to 71° domain walls (black). inset: schematic nanoscale PV measurement setup using a conductive AFM tip as a variable-distance counter electrode for I-V characterization. b) Initial large slope (red line) indicating large PV effect at 109 degree walls. Adapted from Seidel et al [59].

Currently, the overall efficiency of this photovoltaic effect in devices is limited by the conductivity of the bulk bismuth ferrite material. Methods to increase the carrier mobility, as well as inducing the spatially periodic potential in an adjacent material with a lower gap than BFO are possible routes to achieve larger current densities under white light illumination, and more generally, they would demonstrate what the source of periodic potential and the PV current flow can be in different materials. Low band-gap semiconductors with asymmetric electron and hole mobilities are possible candidates to show such an effect. In addition, photoelectrochemic effects at domain walls are a possible further interesting route, e.g. for applications in water splitting. [74]

The spatial and temporal evolution of photoinduced charge generation and carrier separation in heteroepitaxial $BiFeO_3$ thin films was measured with Kelvin probe and piezoresponse force microscopy. [75] Contributions from the self-poled and ferroelectric polarization charge were identified from the time evolution of the correlated surface potential and ferroelectric polarization in both, films as grown and after poling, and at different stages and intensities of optical illumination. Variations in the surface potential with bias voltage, switching history, and illumination intensity were investigated. It was shown that both bulk ferroelectric photovoltaic and the domain wall offset potential mechanisms contribute to the photogenerated charge. Polycrystalline materials, [76] 2-D interfaces, [77, 78, 79, 80] and 1-D ferroelectric nanostructures have been explored for enhanced photovoltaic responses as well, [81, 82] in addition to nanoscale enhancements of ferroelectric photovoltaic effects at metal nano-tips. [83]

Photodetection characteristics in leaky ferroelectric BFO thin films have been investigated. [84] Nano-optical approaches have also been reported for the visualization of n-p junctions formed by oxygen vacancy motion under the application of an electric field in a Ca-doped BiFeO3 thin film through spatially resolved scanning photocurrent mapping. [85] The photocurrent mapping, in conjunction with the spectroscopic approach, provides clues to local electronic structures and defect levels associated with oxygen vacancies.

Steady-state short-circuit photocurrents have been measured in polycrystalline $Pb(ZrTi)O_3$(PZT) films with Pb excess. It was found that the photocurrent is directed opposite to ferroelectric polarization due to PZT grains and semiconductor PbO phases segregated on PZT grain boundaries during the PZT formation. In such a medium, the current across the film flows through nanoscale channels formed by the PbO phase, where an electric field is generated by uncompensated polarization charge of PZT grains. A dependence of photocurrent on polarization was found in this system. [86] This concept has been extended to the inclusion of silver oxide nanoparticles. [87]

A novel approach was demonstrated to synthesize perovskite nanomaterials in an aqueous system using genetically engineered viruses for nano-photovoltaics. [88] $SrTiO_3$ and $BiFeO_3$ nanoparticles were successfully templated on viruses, achieving small particle size and high crystallinity compared to conventional techniques extending the scope of biotemplated synthesis to ternary metal oxides as well as more complex materials. These virus-templated perovskite nanomaterials have shown photocatalytic and photovoltaic performance.

In summary, nanoscale photovoltaic effects in oxides form an exciting and growing field of interest in functional materials. With the current developments surrounding domain walls and topological defects there are many remaining questions and new areas to be explored. The intrinsic dynamics of domain walls and other topological defects are expected not only to influence future theoretical and experimental interpretations of the electronic phenomena, but also pose a possibility to find unique properties of multiferroic domain walls, influenced by order parameter coupling and localized secondary order parameters. [52] Of obvious future interest is the question of what sets the limits to photovoltaic behavior at walls. It might be possible to "design" the topological structure of the domain wall to controllably induce electronic phase transitions within the wall arising from the correlated electron nature. Oxide photovoltaics, particularly with nanoscale ferroelectrics and multiferroics, thus might hold significant potential for future nanotechnology applications. [89]

References

[1] J. Seidel, Domain walls as nanoscale functional elements, *J. Phys. Chem. Lett.* 3, 2905 (2012).

[2] H. Hung, Ferroelectric photovoltaics, *Nature Photonics* 4, 134 (2010).

[3] J. Kreisel, M. Alexe and P. A. Thomas, A photoferroelectric material is more than the sum of its parts, *Nature Mater.* 11, 260 (2012).

[4] J. Seidel, G. Singh-Bhalla, Q. He, et al., Domain wall functionality in BiFeO3, *Phase Trans.* 86, 53 (2013).

[5] Y.-P. Chiu, Y.-T. Chen, B.-C. Huang, et al., Atomic-scale evolution of local electronic structure across multiferroic domain walls, *Advanced Materials* 23, 1530 (2011).

[6] S.-Y. Yang, J. Seidel, S. J. Byrnes, et al., Above-bandgap voltages from ferroelectric photovoltaic devices, *Nature Nanotech.* 5, 143 (2010).

[7] Seidel J., P. Maksymovych, A. J. Katan, et al., Domain wall conductivity in La-Doped $BiFeO_3$, *Phys. Rev. Lett.* 105, 197603 (2010).

[8] J. Seidel, D. Fu, S.-Y. Yang, E. Alarcòn-Lladò, J. Wu, R. Ramesh, and J. W. Ager, *Phys. Rev. Lett.* 107, 126805 (2011).

[9] J. Seidel, S.-Y. Yang, E. Alarcòn-Lladò, J. W. Ager, and R. Ramesh, Nanoscale probing of high photovoltages at 109° domain walls, *Ferroelectrics* 433, 123 (2012).

[10] M. Imada, A. Fujimori, andY. Tokura, *Rev. Mod. Phys.* 70 (1998), pp. 1039–1263.

[11] S. B. Ogale, Thin Films and Heterostructures for Oxide Electronics, Springer, New York, 2005.

[12] A. Ohtomo, D. A. Muller, J. L Grazul, and H. Y.Wang, *Nature* 419 (2002), pp. 378–380.

[13] Dagotto, E. When oxides meet face to face. *Science* 318, 1076−1077 (2007).

[14] Mannhart, J. & Schlom, D. G. Oxide interfaces – an opportunity for electronics. *Science* 327, 1607–1611 (2010).

[15] Yamada, H. et al. Engineered interface of magnetic oxides. *Science* 395, 646−648 (2004).

[16] B. I. Sturman and V. M. Fridkin, *The photovoltaic and Photorefractive Effects in Noncentrosymmetric Materials*, Gordon and Breach, Amsterdam 1992.

[17] A. Prokhorov, *Ferroelectric Crystals for Laser Radiation Control*, Taylor & Francis 1990.

[18] M. E. Lines, A. M. Glass, *Principles and Applications of Ferroelectrics and Related Materials*, Oxford University Press 2001.

[19] S. B. Ogale, T. V. Venkatesan, and M. Blamire, *Functional Metal Oxides: New Science and Novel Applications* Wiley-VCH, Weinheim, 2013.

[20] P. Ferraro, S. Grilli, P. De Natale, *Ferroelectric Crystals for Photonic Applications: Including Nanoscale Fabrication and Characterization Techniques*, Springer Series in Materials Science 2009.

[21] S. Gevorgian, *Ferroelectrics in Microwave Devices, Circuits and Systems: Physics, Modeling, Fabrication and Measurements*, Springer 2010.

[22] V. M. Fridkin, *Photoferroelectrics,* Springer 2011.

[23] P. Günter and P. Huignard, *Photorefractive materials and their applications*, Springer 2010.

[24] D. Ginley, M. A. Green, and R. Collins, *Mater. Res. Soc. Bull.*33, 355 (2008).

[25] I. Gur, N. A. Fromer, M. L. Geier, and A. P. Alivisatos, *Science*310, 462 (2005).

[26] B. O'Regan and M. Grätzel, *Nature* 353, 737 (1991).

[27] B. Goldstein and L. Pensak, *J. Appl. Phys.* 30, 155 (1959).

[28] P. S. Brody and F. Crowne, *J. Electron. Mater.* 4, 955 (1975).

[29] A. M. Glass, D. von der Linde, T. J. Negran, *Appl. Phys. Lett.* 25, 233 (1974).

[30] V. M. Fridkin, *Crystallogr. Rep.* 46, 654 (2001).

[31] S.-Y. Yang, J. Seidel, S. J. Byrnes, P. Shafer, C.-H. Yang, M. D. Rossell. P. Yu, Y.-H. Chu, J. F. Scott, J. W. Ager III, L. W. Martin, and R. Ramesh, Above-bandgap voltages from ferroelectric photovoltaic devices, *Nature Nanotech*. 5 (2010), pp. 143-147.

[32] W. Shockley and H. J. Queisser, *J. Appl. Phys.* 32, 510 (1975).

[33] Zubko, P., Gariglio, S., Gabay, M., Ghosez, P. & Triscone, J.-M. Interface physics in complex oxide heterostructures. *Annu. Rev. Cond. Mat. Phys.* 2, 141–165 (2011).

[34] Heber, J., Enter the oxides, *Nature* 459, 28-30 (2009).

[35] Seidel et al., submitted.

[36] Zhang, J. X., Xiang, B., He. Q., Large field-induced strains in a lead-free piezoelectric material, *Nature Nanotechnol*. 6, 98 (2011).

[37] Eng L. M., Nanoscale domain engineering and characterization of ferroelectric domains, *Nanotechnology* 10, 405 (1999).

[38] Kalinin, S. V., et al., *Rep. Prog. Phys*. 73 (2010) 056502.

[39] Gruverman, A., Rodriguez, B. J., Dehoff, C., Waldrep, J. D., Kingon, A. I., Nemanich, R. J., Cross, J. S., "Direct studies of domain switching dynamics in thin film ferroelectric capacitors" *Appl. Phys. Lett*. 87, 082902 (2005).

[40] Rodriguez, B. J., Jesse, S., Baddorf, A. P., Zhao, T., Chu, Y. H., Ramesh, R., Eliseev, E. A., Morozovska, A. N., Kalinin, S. V., Spatially Resolved Mapping of Polarization Switching Behavior in Nanoscale Ferroelectrics, *Nanotechnology* 18, 405701 (2007).

[41] Jungk T., A. Hoffmann, and E. Soergel, "Impact of elasticity on the piezoresponse of adjacent ferroelectric domains investigated by scanning force microscopy" *J. Appl. Phys*. 102, 084102 (2007).

[42] Yang C.-H., Seidel J., Kim S. Y., Rossen P. B., Yu P., Gajek M., Chu Y.-H., Martin L. W., Holcomb M. B., He Q., Maksymovych P., Balke N., Kalinin S. V., Baddorf A. P., Basu S. R., Scullin M. L., and Ramesh R., Electric modulation of conduction in multiferroic Ca-doped BifeO3 films, *Nature Materials* 8, 485 (2009)

[43] Seidel, J., et al., Prominent electrochromism through vacancy-order melting, *Nature Comm*. 3, 799 (2011).

[44] Balke, N.; Jesse, S.; Morozovska, A.; Eliseev, E.; Chung, D.; Kim, Y.; Adamczyk, L.; and Garcia, R., Nanoscale mapping of ion diffusion in a lithium-ion battery cathode, *Nature Nanotechnology* **5**, 749-754 (2010).

[45] A. Kumar, F. Ciucci, A. N. Morozovska, S. V. Kalinin, S. Jesse, *Nature Chemistry* 3, 707–713 (2011)

[46] Rodriguez B J, Chu Y H, Ramesh R and Kalinin S V 2008 *Appl. Phys. Lett*. 93 142901

[47] Watanabe Y 2007 *Ferroelectrics* 349 190.

[48] Seidel J., L.W. Martin, Q. He, Q. Zhan, Y.-H. Chu, A. Rother, M. E. Hawkridge, P. Maksymovych, P. Yu, M. Gajek, N. Balke, S. V. Kalinin, S. Gemming, F. Wang, G. Catalan, J. F. Scott, N. A. Spaldin, J. Orenstein and R. Ramesh, "Conduction at domain walls in oxide multiferroics", *Nature Materials* 8, 229 (2009).

[49] Seidel J., P. Maksymovych, A. J. Katan, Y. Batra, Q. He, A. P. Baddorf, S. V. Kalinin, C.-H. Yang, J.-C. Yang, Y.-H. Chu, E. K. H. Salje, H. Wormeester, M. Salmeron, and R. Ramesh, Domain wall conductivity in La-doped BiFeO3, *Physical Review Letters* 105, 197603 (2010)

[50] W. Fan, J. Cao, J. Seidel, Y. Gu, J. W. Yim, C. Barrett, K. M. Yu, J. Ji, R. Ramesh, L. Q. Chen, and J. Wu, Large kinetic asymmetry in the metal-insulator transition nucleated at localized and extended defects, *Physical Review* B 83, 235102 (2011)

[51] Wiessner A., J. Kirschner, G. Schafer, and Th. Berghaus, Design considerations and performance of a combined scanning tunneling and scanning electron microscope, *Rev. Sci. Instrum*. 68, 3790 (1997).

[52] Yang B., N. J. Park, B. I. Seo, Y. H. Oh, S. J. Kim, S. K. Hong, S. S. Lee, and Y. J. Park, *Appl. Phys. Lett*. 87, 062902 (2005).

[53] Garcia R. E., B. D. Huey, and J. E. Blendell, *J. Appl. Phys*. 100, 064105 (2006).

[54] Chiu Ya-Ping, Yu-Ting Chen, Bo-Chao Huang, Min-Chuan Shih, Jan-Chi Yang, Q. He, Chen-Wei Liang, J. Seidel, Yi-Chun Chen, R. Ramesh, Ying-Hao Chu, The Evolution of Local Electronic Structure across Multiferroic Domain Walls, *Adv. Mat*. 23, 1530 (2011).

[55] Lubk, A., et al., Evidence of sharp and diffuse domain walls in $BiFeO_3$ by means of unit-cell-wise strain and polarization maps obtained with high resolution scanning transmission electron microscopy, *Phys. Rev. Lett.* 109, 047601 (2012).

[56] Farokhipoor S., and Noheda, B., Conduction through 71° Domain Walls in $BiFeO_3$ Thin Films, *Phys. Rev. Lett*. 107, 127601 (2011).

[57] J. Seidel, G. Singh-Bhalla,Q. He, S.-Y. Yang, Y.-H. Chu, and R. Ramesh, Domain wall functionality in BiFeO3, *Phase Transitions*, DOI:10.1080/01411594.2012.695371.

[58] Choi, T. et al. Insulating interlocked ferroelectric and structural antiphase domain walls in multiferroic $YMnO_3$. *Nature Mater*. 9, 253−258 (2010).

[59] D. Meier, J. Seidel, A. Cano, K. Delaney, Y. Kumagai, M. Mostovoy, N. A. Spaldin, R. Ramesh, M. Fiebig, Anisotropic conductance at improper ferroelectric domain walls, *Nature Mater*. 11, 284 (2012).

[60] Lubk A., S. Gemming, and N. A. Spaldin, "First-principles study of ferroelectric domain walls in multiferroic bismuth ferrite" *Phys. Rev. B* 80, 104110 (2009).

[61] Meyer B. and D. Vanderbilt, "Ab initio study of ferroelectric domain walls in PbTiO3" *Phys. Rev. B* 65, 104111 (2002).

[62] Hong, L., Soh, A.K., Du, Q.G., Li, J.Y., Interaction of O vacancies and domain structures in single crystal BaTiO3: Two-dimensional ferroelectric model, Phys. Rev. B 77, (2008) 094104.

[63] A. Y. Borisevich et al., *ACS Nano* 4, 6071 (2010).

[64] Guyonnet, J., Gaponenko, I., Gariglio, S., Paruch, P., Conduction at Domain Walls in Insulating $Pb(Zr_{0.2}Ti_{0.8})O_3$ *Thin Films, Advanced Materials*, 23, 5377 (2011).

[65] Schröder, M., Conducting Domain Walls in Lithium Niobate Single Crystals, *Adv. Funct. Mater*. 22, 3936 (2012).

[66] Zhao T., A. Scholl, F. Zavaliche1, K. Lee, M. Barry, A. Doran, M. P. Cruz, Y. H. Chu, C. Ederer, N. A. Spaldin, R. R. Das, D. M. Kim, S. H. Baek, C. B. Eom and R. Ramesh, Electrical control of antiferromagnetic domains in multiferroic BiFeO3 films at room temperature, *Nature Materials* 5, 823 (2006).

[67] Lebeugle D., D. Colson, A. Forget, M. Viret, A. M. Bataille, and A. Gukasov, Electric-Field-Induced Spin Flop in BiFeO3 Single Crystals at Room Temperature *Phys. Rev. Lett.* 100, 227602 (2008).

[68] Idlis, B.G., Usmanov, M.S., Effect of domain structure on the energy spectrum of narrow-gap ferroelectric semiconductors, *Pis'ma Zh. Eksp. Teor. Fiz.* 56, 5, (1992) 268 - 271.

[69] Xiao, Y., Shenoy, V. B., Bhattacharya, K., Depletion Layers and Domain Walls in Semiconducting Ferroelectric Thin Films, *Phys. Rev. Lett.* 95, (2005) 247603.

[70] Gureev, T. M. Y., Tagantsev, A. K., Setter, N., Structure and energy of charged domain walls in ferroelectrics, 18th IEEE ISAF Proceedings (2009).

[71] Yang S.-Y., J. Seidel, S. J. Byrnes, P. Shafer, C.-H. Yang, M. D. Rossell. P. Yu, Y.-H. Chu, J. F. Scott, J. W. Ager III, L. W. Martin, and R. Ramesh, Abovebandgap voltages from ferroelectric photovoltaic devices, *Nature Nanotechnology* 5,143 (2010).

[72] J. Seidel, D. Fu, S.-Y. Yang, E. Alarcòn-Lladò, J. Wu, R. Ramesh, and J. W. Ager; Efficient photovoltaic current generation at ferroelectric domain walls, *Physical Review Letters* 107, 126805 (2011).

[73] J. Seidel, S.-Y. Yang, E. Alarcòn-Lladò, J. W. Ager, and R. Ramesh, Nanoscale probing of high photovoltages at 109° domain walls, *Ferroelectrics* 433, 123, (2012).

[74] Kudo, Akihiko, Y. Miseki, Heterogeneous photocatalyst materials for water splitting, *Chem. Soc. Rev*., 2009, 38, 253.

[75] Feng Yan, Guannan Chen, Li Lu, and Jonathan E. Spanier, Dynamics of Photogenerated Surface Charge on BiFeO3 Films ACS Nano 6, 2353 (2012).

[76] Bin Chen,Mi Li, Yiwei Liu, Zhenghu Zuo, Fei Zhuge, Qing-Feng Zhan and Run-Wei Li, Effect of top electrodes on photovoltaic properties of polycrystalline BiFeO3 based thin film capacitors, *Nanotechnology* 22 (2011) 195201.

[77] T Choi, L Jiang, S Lee, T Egamiand H N Lee, High rectification and photovoltaic effect in oxide nano-junctions, *New Journal of Physics* 14 (2012) 093056.

[78] L. Pintilie, C. Dragoi, and I. Pintilie, Interface controlled photovoltaic effect in epitaxial Pb(Zr,Ti)O3 films with tetragonal structure, *J. Appl. Phys.* 110, 044105 (2011).

[79] T. L. Qu, Y. G. Zhao, D. Xie, J. P. Shi, Q. P. Chen et al., Resistance switching and white-light photovoltaic effects in BiFeO3/Nb–SrTiO3 heterojunctions, *Appl. Phys. Lett*. 98, 173507 (2011).

[80] C-S Tu, C-M Hung, V H Schmidt, R R Chien, M-D Jiang and J Anthoninappen, The origin of photovoltaic responses in BiFeO3 multiferroic ceramics *J. Phys.: Condens. Matter*. 24 (2012) 495902

[81] Gobinda Gopal Khan, Rajasree Das, Nillohit Mukherjee, and Kalyan Mandal, *Phys. Status Solidi RRL* **6**, 312–314 (2012).
Effect of metal doping on highly efficient photovoltaics and switchable photovoltage in bismuth ferrite nanotubes.

[82] Xuehui Zhang, Haiyang Liu, Bin Zheng, Yuanhua Lin, Deping Liu, and Ce-Wen Nan, Photocatalytic and Magnetic Behaviors Observed in BiFeO3 Nanofibers by Electrospinning, *Journal of Nanomaterials*, Volume 2013, 917948 (2013).

[83] Marin Alexe1 & Dietrich Hesse, Tip-enhanced photovoltaic effects in bismuth ferrite, *Nature Communications* 2, 256 (2011).

[84] Won-Mo Lee, Ji Ho Sung, Kanghyun Chu, Xavier Moya, Donghun Lee, Cheol-Joo Kim, Neil D. Mathur, S.-W. Cheong, C.-H. Yang, and Moon-Ho Jo, Spatially Resolved Photodetection in Leaky Ferroelectric BiFeO3, *Adv. Mater*. 2012, 24, OP49–OP53.

[85] K. Kamala Bharathi, Won-Mo Lee, Ji Ho Sung, Ji Soo Lim, Seung Jin Kim, Kanghyun Chu, Jung Won Park, Jong Hyun Song, Moon-Ho Jo, and Chan-Ho Yang, Detection of electrically formed photosensitive area in Ca-doped BiFeO3 thin films, *Appl. Phys. Lett*. 102, 012908 (2013).

[86] Delimova, L. A.), Yuferev, V. S, Grekhov, I. V., Afanasjev, P. V., Kramar, G. P., Petrov, A. A., Afanasjev, V. P., Origin of photoresponse in heterophase ferroelectric Pt/Pb(ZrTi)O-3/Ir capacitors, *Appl. Phys. Lett.* 91, 112907 (2007)

[87] Xiaoluan Yang, Xiaodong Su, Mingrong Shen, Fengang Zheng, Yu Xin, Lu Zhang, Muchuan Hua, Yajie Chen, and V. G. Harris, Enhancement of Photocurrent in Ferroelectric Films via the Incorporation of Narrow Bandgap Nanoparticles, *Adv. Mater*. 2012, 24, 1202–1208.

[88] Nurxat Nuraje, Xiangnan Dang, Jifa Qi, Mark A. Allen, Yu Lei, and Angela M. Belcher, Biotemplated Synthesis of Perovskite Nanomaterials for Solar Energy Conversion, *Adv. Mater*. 2012, 24, 2885–2889.

[89] G. Catalan et al., *Rev. Mod. Phys*. 84, 119 (2012).

INDEX

D

E

F

G

H

I

N

O

P

S

T

U

V